MODERN COMBAT SHIPS 2
'Invincible'
Class

D1743095

Above:
Amongst the main differences visible from the outside,
***Illustrious* has the forward Phalanx CIWS farther**
forward and the deck crane farther aft, than *Invincible*.
The life rafts are grouped slightly differently and the
bridge top has been redesigned to carry satellite
communications gear. In this aerial view of the ship
returning to Portsmouth after its 1982 deployment to
the South Atlantic, the Sea Harrier 'runway' and the
helicopter landing 'spots' are easily visible.
HMS Illustrious

MODERN COMBAT SHIPS 2

'Invincible' Class

Paul Beaver

LONDON
IAN ALLAN LTD

Contents

Unless otherwise credited, the photographs in this book are British Crown Copyright

Front cover:
HMS *Invincible* executes a tight turn to starboard. *RN*

Back cover, top:
***Illustrious* arriving at Portsmouth on 28 June 1982 gave watchers something of a surprise when they saw the Phalanx CIWS systems and other armaments modifications.** *Robin Walker*

Back cover, bottom:
***Ark Royal* building. Unlike *Invincible*, *Ark Royal's* design incorporated the ski-jump ramp from the outset.** *Swan Hunters*

First published 1984

ISBN 0 7110 1386 1

Published by Ian Allan Ltd, Shepperton, Surrey; and printed by Ian Allan Printing Ltd at their works at Coombelands in Runnymede, England.

Preface

There cannot have been a class of British warship which has been so controversial as the 'Invincible' class since the 'Dreadnoughts' at the turn of the 19th century. Certainly, no type of warship has been used so often as a political pawn. Despite both of these observations, the 'Invincible' class has been proved in action in a limited but testing war situation many miles from home ports. Although there are destined to be only three of the class they will make a tremendous contribution to the Royal Navy, and through it, to the North Atlantic Treaty Organisation. They have shown their anti-submarine capability and this, almost above all else, is the need of the Western navies in the remaining years of this century and the first decade of the next.

Throughout the book I have referred to the class' current role as that of a light aircraft carrier which should really have the abbreviation CVL; however, to conform with current naval thinking the three ships, *Invincible*, *Illustrious* and *Ark Royal*, are rated as anti-submarine warfare aircraft carriers, or CVSs. It is the latter abbreviation which I have used. Furthermore, I have occasionally found it necessary to differentiate between the last conventional aircraft carrier in the Royal Navy, *Ark Royal*, and the present 'Invincible' class warship of the same name. Where there is a risk of confusion, I have noted the former as *Ark Royal* (IV) — being the fourth ship to bear the name in the RN — and the latter as *Ark Royal* (V). The use of Roman numerals is not official practice but a recognised method of showing the chronology of a warship's name.

Acknowledgements

It is impossible of course to write such a book as this without official assistance, and even backing from the Royal Navy; however, it must be stressed that this is very much my work and does not reflect any official opinion. Nevertheless, I am most grateful to the Director of Public Relations (Navy), Capt Ian Sutherland, and his staff for support, advice and considerable assistance.

I am particularly grateful to Capt The Honourable Nicholas Hill-Norton who was kind enough to invite me aboard *Invincible* during Exercise 'Rough Diamond' in the English Channel, and again to take passage in his magnificent ship between Australia and New Zealand during the 'Orient Express' deployment. Both these facilities allowed me to view at first hand the workings of what is the largest class of warship to have been designed and built in the UK since 1945. Rear Adm Jeremy Black DSO, MBE also gave me an insight to the problems, and joys, of command of the ship during the Falklands conflict, and was gracious enough to allow me certain privileges during 'Orient Express'.

To portray accurately the background to the design, I was very lucky to be able to draw upon the thoughts and advice of Adm Sir Anthony Griffin GCB, a former Controller of the Navy, carrier captain and chairman of British Shipbuilders.

I would like to extend my thanks to the following who have greatly assisted the project: Cdr Martin Nestor, Lt Cdr Gerry Hunt and Lts Alaistair McLaren, Tony Wilkinson, Dick Ayres and Garrick Beats in *Invincible*. In industry, Sue Elfring (Rolls-Royce), J. M. Pullen, W. G. Clouter, Mike Smith, Dennis Shadbolt (British Shipbuilders), John McNally (BMARC), S. C. Platten (David Brown) and Mike Farlam (Westland) all contributed with technical information and photographs. Laurie Phillips and Michael Hill were also most helpful at Commander-in-Chief Fleet's headquarters, Northwood, as were Cdr Paul Madge, Lt Cdr Brian Morgan and Leading Airman (Photographer) Nigel Thomas at Yeovilton. At Culdrose, Second Officer S. J. Eagles WRNS provided some sought after data on flight deck vehicles and Cdr G. Sullivan, the SIO in *Illustrious*, was able to fill in all the gaps regarding his ship's movements and equipment. In the above, it has not been possible to thank all of 801 and 820 Squadrons by name, nor the detachment from 817 Squadron, Royal Australian Navy, but I am just as grateful to them.

Paul Beaver

1 Background — The First of a New Breed

The aircraft carrier is a relatively new breed of warship, born in the latter years of World War 1 to solve the problem of putting aircraft to sea in usable numbers. It was not until 1924 that the first such warship designed from the keel up came into service, although the converted collier, *Ark Royal* (II), was the first warship to carry aircraft to war. By the end of World War 2 the aircraft carrier, particularly the fleet carrier, had replaced the battleship as the capital ship of the world's two major navies — the United States Navy (USN) and the Royal Navy (RN). The immediate postwar years saw a rapid decline in the number of warships declared operational and it was not until the Korean War (1950-53) that warship building and modernisation began again in earnest. Britain in particular had the problem of a widely dispersed empire which was rapidly seeking independence, not always in a democratic and orderly fashion, leaving the UK with a legacy from the war of yet more wars. In

addition the growing Soviet naval strength, particularly in submarine forces, caused the Admiralty to look closely at the design of aircraft carriers for a possible conflict in the North Atlantic. With the advent of jets and helicopters, some radical rethinking was necessary.

The first postwar carrier design study was the modernistic and potentially highly effective CVA-01 (CVA = attack carrier) design with a displacement of 54,000 tons and an overall length of 890ft (271m). Compared to the immediate

Below:

The 'Invincible' class has been given several titles during its brief career with the Royal Navy — light aircraft carrier, helicopter cruiser or through-deck cruiser. Many thought in fact that the ships would never see service; luckily they were wrong. This is *Invincible* **pictured on builders trials in the Clyde — note the markings on the flight deck which indicate the hull shape at the waterline.**
Vickers Shipbuilding & Engineering Ltd

prewar *Ark Royal* (III), this was an increase of 26,000 tons in displacement, but barely 90ft (27m) in length. The extra weight was taken up by the installation of an angled deck capable of taking Mach 2 jets with gross landing weights of more than 50,000lb (22,680kg), as against the average 1938-vintage aeroplane's 8,228lb (3,732kg). The flight deck design, including 'round-the-island' taxiways, represented an increase of 15% usable deck space. The angled deck was destined for a long-stroke steam catapult, as well as one at the bow, and the arrester gear was to be the revolutionary water-spray type.

Politically, however, the cost of the CVA-01 and its two or three sister ships would have been too much for an already strengthened defence budget which, in the jet era and considering the priority given to the British Army of the Rhine, gave the major emphasis to the Royal Air Force and British Army. In addition, the North Atlantic Treaty Organisation (NATO) saw the future RN role primarily as anti-submarine warfare (ASW), a naval specialisation in which the British have excelled since the 1940s. Navies were beginning to develop the idea of carrying a small helicopter in fast frigates to provide anti-submarine striking power beyond the range of normal shipborne systems, the Westland Wasp HAS1 being the British development in this field. The carrier therefore was seen more in the light of being an ASW ship, with an air component for defence, rather than being a large strike carrier like those of the 'Forrestal' class which were being designed for the United States Navy. For this reason and because of the disproportionate cost of the CVA-01 concept, which included special escorts of the 'Bristol' class, to the naval budget as a whole, many Admiralty staff officers began to consider the escort carrier concept again. During World War 2 the use of specially-converted merchant ships as light carriers had turned the tide of the Battle of the Atlantic against the German submarine menace. Hence the use of lighter and smaller escort carriers, with flight decks suitable for large ASW helicopters, was considered. They would also serve as comprehensive ASW command and control centres. At the same time, the Hawker Siddeley company (now part of British Aerospace) was developing the vertical/short take-off and landing (V/STOL) design which has now led to the successful Harrier and Sea Harrier family. No serious study was undertaken and, as can be expected, Fleet Air Arm (FAA) officers who were informed of the ideas were appalled at the thought of losing 'normal' fixed-wing jet aviation.

In 1962, the question was posed by the Vice-Chief of Naval Staff: 'If CVA-01 is not approved, what should be the characteristics of the RN's future fleet?'. His staff came up with the idea of two types of cruiser (14,000 and 17,000 tons) which could carry both large ASW helicopters and V/STOL aircraft, and take advantage of the new breed of missiles for self-defence coming into service. Amongst the purely naval advantages would be the smooth and clear decks with no gear necessary to work aircraft (steam catapults and arrester gear, both being heavy systems, increased topweight, as well as causing inconvenience by their physical presence). Also, the ship would be easier to manoeuvre, thus appealing to the seamanship side of the RN. The studies were completed by November 1962, but then promptly shelved.

By 1966, there had been a change of government to one with a rather narrow approach to foreign and defence policies, including the withdrawal of British forces from East of Suez. Meanwhile the RAF, in the guise of Air Chief Marshal Sir Thomas Pike, had postulated the idea of a fleet of converted merchant ships to act as mobile air bases, as and when required, with the back-up of 'island-hopping' RAF aircraft, using various independent nations' facilities to refuel. In addition the RAF actually appears to have put forward the idea of the FAA being returned to its control — as it had been so disastrously prewar, leading to a lack of equipment which contributed to Britain's near defeat at sea in the early war years.

The 'Pike Ship Fleet' concept was successfully resisted by the RN, and it is interesting to consider the Falklands experience of using converted merchant ships, particularly container vessels, to transport aircraft south and even act as flight decks for operations. The concept has now been taken a little farther with the US Arapaho system being investigated in the *Astronomer* project by the RN. This ship has been converted to carry five Sea King helicopters but needs an incredible 140 containers to carry the necessary support equipment. This type of ship could only back up the existing 'Invincible' class, not take over from them.

The Labour Government of the Day's Defence Secretary, Denis Healey, cancelled the CVA-01 programme in 1966, leaving the RN without any replacement for the ageing *Victorious*, *Ark Royal*, *Eagle* and *Hermes*. The RN had not only lost its future organic air cover for strike and air defence, including airborne early warning (AEW), but also command and control (C&C) facilities for major ASW operations. It was decided however to convert *Ark Royal* to operate Phantoms, and the 'Tiger' class cruisers were refitted to operate helicopters — initially four Westland Wessex HAS3s, later replaced by Westland Sea King

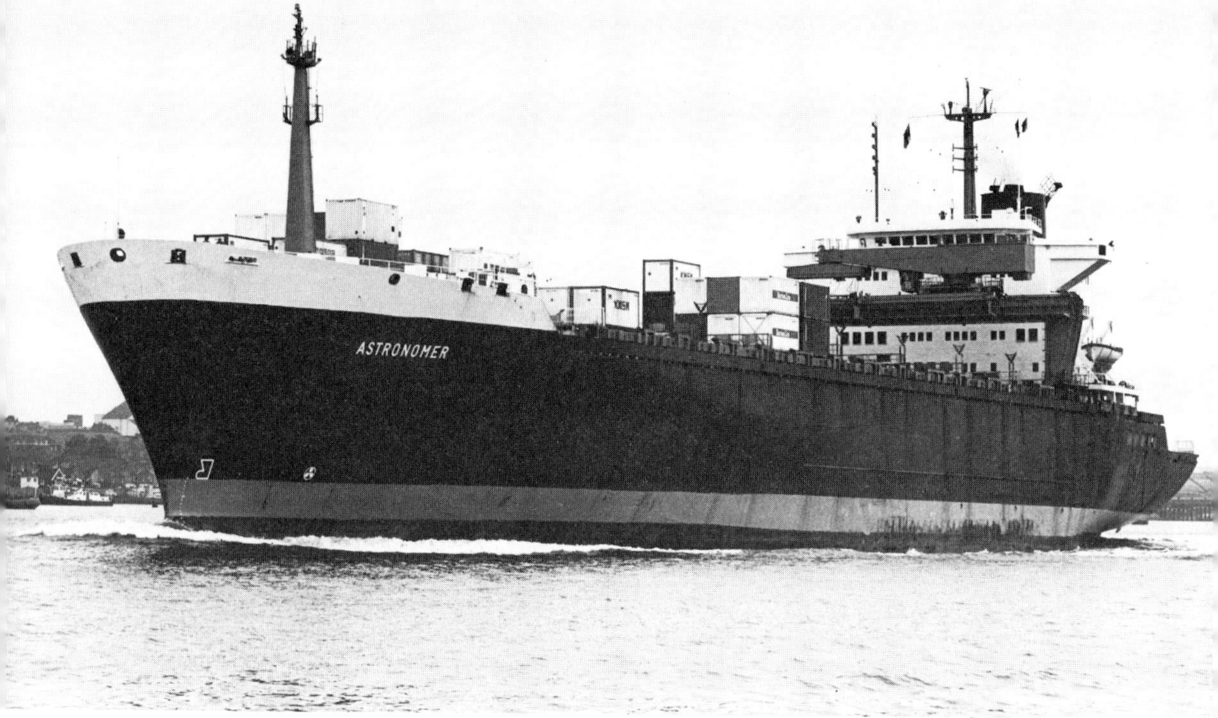
ASTRONOMER

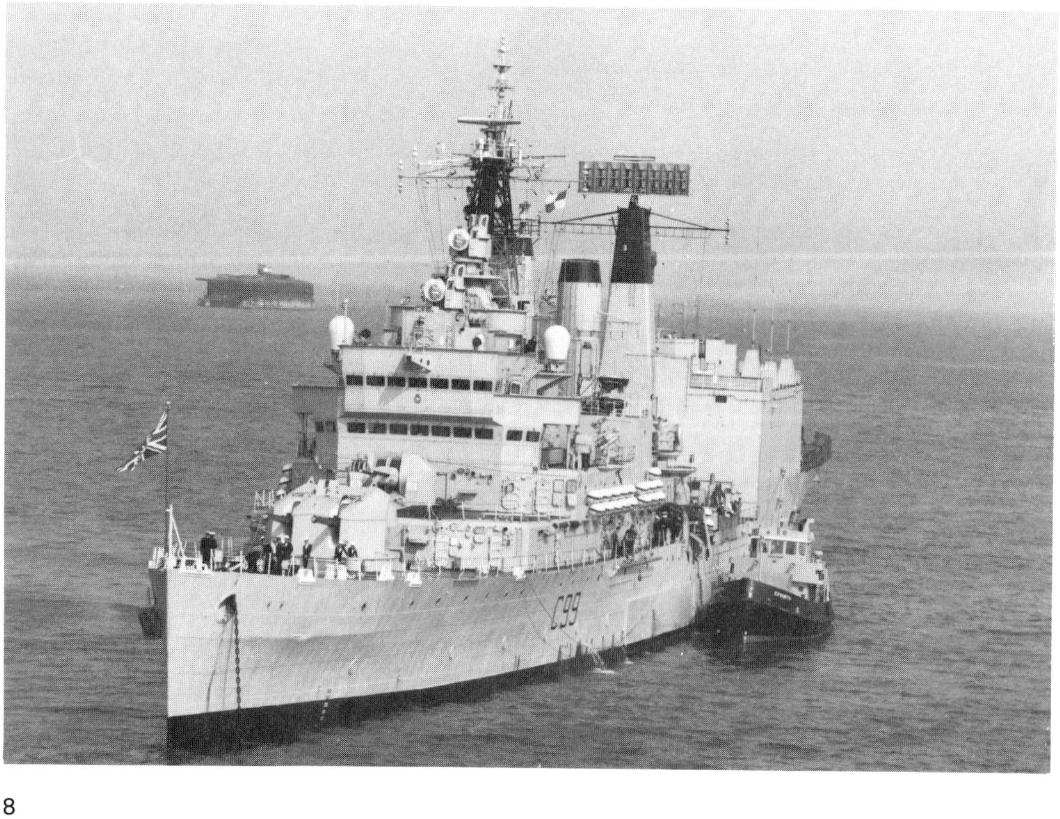
C99

HAS1s — and given some C&C facilities. But there was also a need to replace the 'Tigers'. The provision of any new fixed-wing aircraft, including VSTOL aircraft, would not have been politically tolerable at this time, despite the successful trials carried out aboard *Ark Royal* (IV) in 1963 to prove the use of the P1127 in a maritime environment. The case for VSTOL was probably not helped by FAA's concern about the performance of the proposed P1154 compared to the readily available McDonnell-Douglas Phantom which could be operated from the existing carriers; in the event only *Ark Royal* was modified to take the aircraft. It is ironic that the RAF later chose the Phantom for its standard strike fighter and the naval Buccaneer for its low-level strike aircraft!

The first new design considered was a 12,500-ton command cruiser with a complement of about 1,000 men and facilities to operate six of the American-designed but British-built SH-3 (or Sea King) helicopters. The number of helicopters was small because the hangar deck was proposed in the island superstructure rather than below the flight deck. If a below-decks hangar was required then the cruiser had to be of the through-deck configuration and the displacement increased to 17,500 tons, with the superstructure moved to the conventional starboard side of the deck. This would allow 12 Sea Kings to be carried, three on deck and nine below. In addition the design would have the new GWS30 Sea Dart area air defence missile system and the GWS25 Sea Wolf point defence missile system. Adequate air warning and search radar would be carried, together with

a full suite of command, control and communications (C^3) equipment.

It was inevitable that the latter concept was more acceptable than the smaller ship and it was therefore progressed both by the constructors at Bath and the Admiralty Staff. The CCH concept therefore became the CAH (assault/helicopter cruiser), with an increase in displacement to 19,810 tons incorporating the following important features, which are reflected today in the operational 'Invincible' class design.

The flight deck would be clear and unobstructed, allowing for a rolling launch and recovery giving, for the first time, the possibility of operating VSTOL aircraft. This was kept rather in the dark lest some politician should decide that the RN was trying to get back in to fixed-wing aviation again. Below the flight deck there were to be adequate facilities, especially in the form of workshops around the hangar deck, for the easy maintenance of aircraft. This would be of particular importance in view of the limited numbers which could be carried.

Three major roles were envisaged for the CAH and all of these were tested in operational conditions by the first ship of the type, *Invincible*, during its duties with the South Atlantic carrier battle group (CVBG) in April, May, June, July, August and September 1982.

The first and certainly the most important role was that of providing an operating base for the deployment of the large ASW helicopters of the Sea King, and later, the EH101 type. Large helicopters have to be deployed in groups of squadron strength to be effective and to provide a constant screen of anti-submarine support in advance of any task group. Although designed to lead a NATO ASW task group in the North Atlantic, the CCH/CAH concept of the through-deck cruiser did not perceive the equipping of the ships with their own ASW systems. The sole reliance on the Sea King helicopters, excellent submarine hunters that they are, is a concept not shared by Italy, whose navy is considered quite rightly to be second only to the RN in ASW operations. The Italian Navy has recently completed a light aircraft carrier in a similar concept, albeit rather smaller at 13,370 tons. The *Giuseppe Garibaldi* not only carries close-in guns and missiles, but also six ship-launched ASW torpedo tubes as well. Its hangar can accommodate 12 Agusta-built SH-3D Sea Kings too. The RN on the other

Below:

The main role of the class is to take to sea a large number of the Fleet Air Arm's largest helicopter, the Sea King HAS5. Operating in the Southern Ocean, two Sea Kings of 820 Squadron return from an anti-submarine exercise. *Author*

hand has only allowed for the provision of Type 184 sonar (to be replaced in *Invincible* and *Illustrious* with Type 2016, which is fitted to *Ark Royal* [V]). This hull-mounted sonar does not have the back-up of close range, 'last ditch' torpedo systems.

The second role envisaged for the ships was providing organic air cover to the Fleet, giving a potential for air-to-air combat and surface strike. In fact it became apparent to the FAA in 1970 that it needed to have a stake in the vertical take-off aircraft scene and therefore design studies into the use of the Harrier at sea were pursued. Trials with land-based variants had been carried out in 1969, but it was not until 1971 that a naval staff requirement (NSR) was issued for a navalised variant of the Harrier GR3. Details of the requirement and role will be found in Chapter 6, but suffice to say that it was not until May 1975 that the Sea Harrier was considered to be part of the class' overall design. The idea of providing airborne early warning (AEW) for the ships had not been actively pursued, with trust being placed in the newly developed Type 1022 air warning radar and the use of RAF land-based aircraft, such as the elderly Shackleton and the Nimrod (due to enter service only in 1984!). One design study was carried out by Westland Helicopters in 1967-68, but the cost of putting a radar system into a helicopter, which is relatively unstable and prone to vibration, was deemed 'out of court' on cost grounds. In 1982 a successful marriage of AEW radar and helicopter was carried out, and details can be found in chapter 6.

The third role anticipated for the 'Invincible' class was that of providing full and complete flagship services — the command and control ship. In a modern navy, especially in a modern war scenario of dispersed warships performing different but complementary tasks, it is important that the task group commander can, at all times, keep a firm hand on the proceedings. This is done by monitoring the radar and sonar information available, from all sources, on a centralised 'plot' in the ship's operations room. To fit out a ship for this task is no easy matter as it requires numerous spaces within ship's superstructure to be available, besides a large and well-planned operations room and comprehensive aerial/antennae fit

Below:
Another role given to the 'Invincibles' is that of carrying a Royal Marine Commando. During Exercise 'Rough Diamond' *Invincible* embarked two squadrons of support helicopters for amphibious operations on the south coast of England. Illustrated are two Sea King HC4s from 846 Squadron and, nearest the camera, a Wessex HU5 of 845 Squadron. *Author*

Above:
It was a bleak day when *Invincible* arrived at Portsmouth for official acceptance into the Royal Navy. There was already some controversy about the design, especially concerning the large propellers, one of which had to be replaced very soon after this picture was taken in March 1980. *Author*

Left:
The first commanding officer of *Invincible* was Capt Michael Livesay RN, pictured here on the flight deck of his ship before it sailed for Portsmouth. *RN*

'topside'. In fact, the design process for the actual layout of the island superstructure off the so-called command or through-deck cruiser took this into account. In addition to the standard radar and radio antennae, space had to be made available for the latest satellite communications and navigation equipment. It is interesting to note that when *Illustrious*, the second of the class, entered service in 1982, it was arguably the best-equipped command ship in the world.

Any task group in a hostile situation is likely to become a target for air attack, as indeed *Invincible* and *Hermes* were during their time in the South Atlantic during 1982. In fact, many 'Latins' still believe that the former ship was badly damaged by an air-launched Exocet missile or even sunk by Iron bombs, following an Argentine air strike. Pictures were even released to the Argentine press showing *Invincible* in flames;

unfortunately the prints used were from an official RN portfolio and the pictures were seen to be fakes. However, RN damage control posters have made use of the pictures with the slogan 'Damage Control saved *Invincible*'! It is surprising then, that when the CCH plan was sent to the Admiralty Board for approval in the late 1960s (which was followed by ministerial approval by Lord Carrington, the then Secretary of State for Defence, in 1971), the inclusion of point defence missiles on the fo'c'sle and ships' quarters was deleted for economic reasons, with the feeling that there would be enough Sea Wolf-equipped escorts to deal with this threat, assuming that a CCH/CAH would not venture to sea alone. The Board was nearly proved wrong in the Falklands when only two Sea Wolf-equipped Type 22 frigates were present. Today, it is thought that the development of the vertical-launch Sea Wolf for the Type 23 frigate might lead to the provision of the system in the 'Invincible' class during refit.

The design was given some self-defence of course, with the provision of a single Sea Dart launcher for area air defence. In addition, the provision of the Type 42 destroyers with Sea Dart for task group area air defence was cited as a reasonable explanation for giving a 19,810-ton ship only one defensive system. The command and control of the ship's own fighters and those from other sources, both afloat and shore-based, does allow for an additional vestige of protection, but many believe the class to be under-protected.

The ability of the CCH/CAH design to defend itself from a surface threat was not deeply considered, bearing in mind the modus operandi of the type. However, it is worth noting that both the Sea Dart and the Sea Harrier have surface strike capabilities to a lesser and greater extent respectively. This is an area where it is thought that there will not be much development of the ships' capability in the future.

As a final touch to the design, in 1975-76 amendments were made to allow for the emergency embarkation of a Royal Marine Commando for quick-dash transportation to such areas as the Northern flank of NATO, or other potential trouble spots. This facility was used during the deployment of the South Atlantic Task Force when *Invincible* took Royal Marines part of the way to the Falklands.

Although the through-deck cruiser design was drawn from the 'Tiger' class cruisers, two of these had been converted to carry helicopters (*Tiger* and *Blake*), and lessons learned from them were of great importance and led to a totally new conception of how to take helicopters and later STOVL — Short Take-Off and Vertical Landing — aircraft to sea.

By the time that *Ark Royal* (IV) was decom-missioned in 1978, the CAH concept had grown to that of a light aircraft carrier (CVL), which would have the capability of carrying nine Sea King HAS2 helicopters and five Sea Harrier STOVL jets in a normal complement; this could be increased to 12 Sea Kings and eight Sea Harriers in time of war. The use of the term aircraft carrier for the design is still not universally accepted, although the famous *Jane's Fighting Ships* now gives the 'Invincible' class that designation. Those who perhaps still favour RAF domination of the FAA still have an inclination to call the design an escort cruiser. It could of course be rated an 'aircraft-carrying cruiser'. For all intents and purposes, however, ships' operation is as light aircraft carriers.

It was as an escort cruiser design that the Admiralty naming committee had to provide a name for the lead ship in the class. This first name carries tremendous weight and prestige, especially in a service with such tradition and history as the Royal Navy. It would be almost unthinkable to invent new names or to use those of a previous class which did not approximately correspond to the role or specification of the new ships. As an escort cruiser it would have been politically unpopular to have named the design after an aircraft carrier class of the immediate past, as was suggested by several sources. It was therefore necessary to look back at successful battlecruisers and flagships as the design was certainly closely related to those design philosophies. The name which immediately sprang to mind was *Invincible*, being Adm Sturdee's flagship at the Battle of the Falkland Islands in 1914, and a battlecruiser to boot. Another battlecruiser, *Lion*, might well have been second choice and although the name had been used for one of the 'Tiger' class cruisers of the postwar era, it had not been converted into a helicopter cruiser, having been broken up at Inverkeithing in April 1975. *Lion* also had Battle of Jutland connections. When it came to naming the second of what is now the 'Invincible' class the political situation had changed and an aircraft carrier name could be used. That of the famous World War 2 carrier *Illustrious* was chosen, having a similar sort of message in the name. The third ship ordered was laid down on 14 December 1978; called *Ark Royal*, it was launched during a period of national interest in the proposals to keep *Ark Royal* (IV) as a museum ship, and was named, one suspects, partly to humour public opinion. The name chosen also gave more feeling to the association with the fourth *Ark Royal* and HM The Queen Mother's connection with both ships, as patron. The original name chosen by the RN, in relation to *Illustrious* and the line of World War 2 carriers, was *Indomitable*.

2 Construction

The first of the 'Invincible' class — the nameship itself — was constructed at the Barrow-in-Furness shipyard of Vickers Shipbuilding & Engineering Ltd, a subsidiary company of the nationalised British Shipbuilders. Vickers began the design-assistance contract for ship weapon systems engineering in the early 1970s and the build contract was authorised on 17 April 1973. The building of the ship required the skills and services of over 3,000 men and women at Barrow alone, but unfortunately the launch date was postponed several times, until 3 May 1977, when HM The Queen named the ship. *Invincible* commenced Contractor's Sea Trials in the Irish Sea and the Clyde Approaches during April 1979 and was accepted into the RN at Portsmouth on 19 March 1980.

Illustrious was ordered from Swan Hunters, also part of the British Shipbuilders warship group, in May 1976, being laid down on 7 October of that year. The ship was launched at Hebburn on 1 December 1978, and two weeks later the third of the class, now named *Ark Royal*, was laid down in the same yard. The sea trials for *Illustrious* were commenced in early 1981, although completion was speeded up due to the situation in the South Atlantic during early 1982. Chapter 7 has further details about the ships' subsequent careers and the part they played in the Falklands campaign.

Right:
The first ship of a class is always rather longer in building than later ships. This is not surprising for the 'Invincibles' because although there had been considerable planning and computer-related studies, it is the first time that anything on this scale had been completed by a British yard since World War 2. *Invincible* **was laid down on 20 July 1973 and launched on 3 May 1977; the ship was completing fitting out when this picture was taken at Barrow-in-Furness in 1979.** *VSEL*

Below right:
By March 1979 the ship was beginning the first of several months of trials in the Irish Sea and Clyde to test all its systems and equipment. Many of the features were new and untried. This view shows *Invincible* **during turning trials.** *VSEL*

Design Task Phases: *Invincible*

Date	Company	Design Task
November 1969	Rolls-Royce	Olympus Engine Ducting
February 1970	Y-ARD	Propulsion Machinery Computer Simulation
June 1970	Y-ARD	Machinery Specification and Guidance Drawings
September 1970	MacTaggart Scott	Aircraft Lift Design Study
December 1970	Rolls-Royce	Shore Test of Main Machinery
March 1971	AS Computers	Structural Analysis Calculations
October 1971	Rolls-Royce, Vospers and Hawker Siddeley	Main Machinery Control System
April 1972	Vickers Shipbuilding	Preparation of Working Drawings (over 30,000)
April 1972	Vickers Shipbuilding	1/10th Scale Machinery Space Models
June 1972	Y-ARD	Reliability Study
August 1972	Vickers Shipbuilding	Preparation of QC Documentation Test Forms
August 1972	Vickers Shipbuilding	Specification and Procurement of Long Lead Equipment
May 1973	Vickers Shipbuilding	Support Tasks
May 1978	Vickers Shipbuilding	First-of-Class Trials

Note: This information was provided by Vickers Shipbuilding Group and demonstrates the vast amount of work necessary to design and build a major warship.

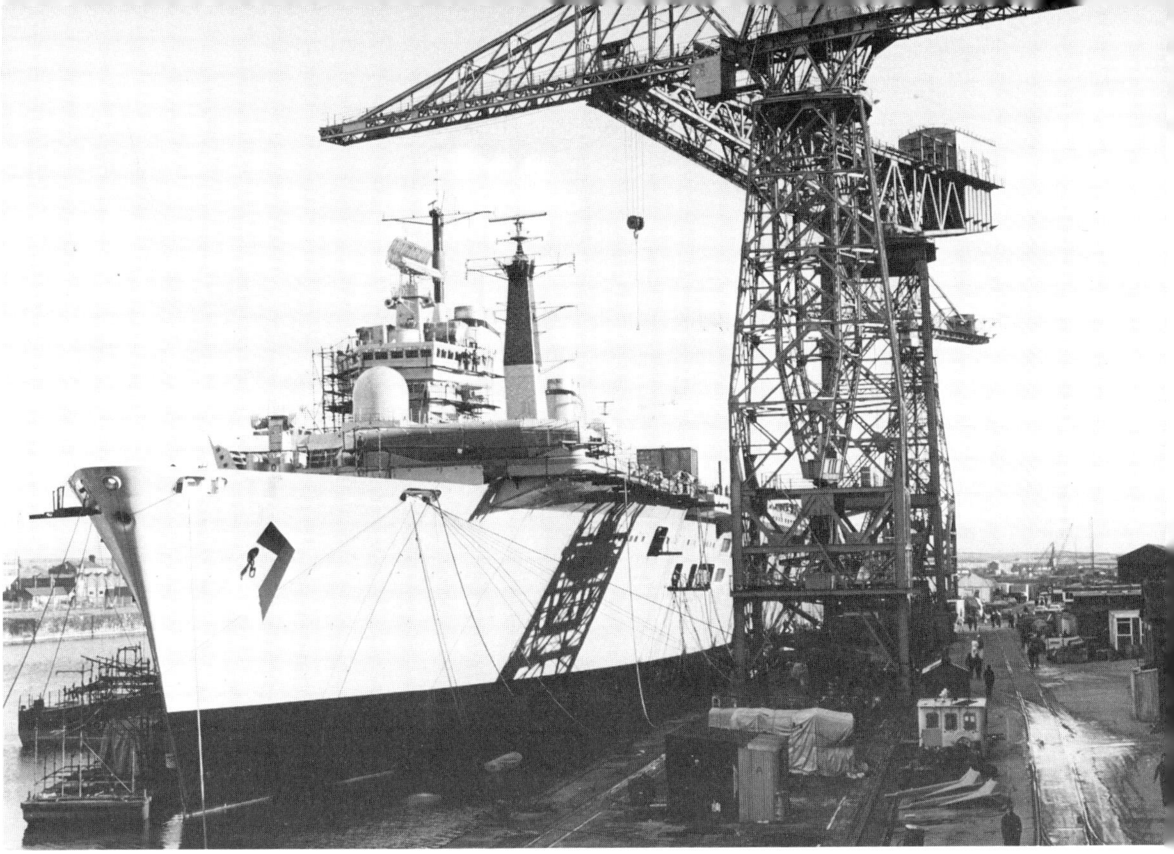

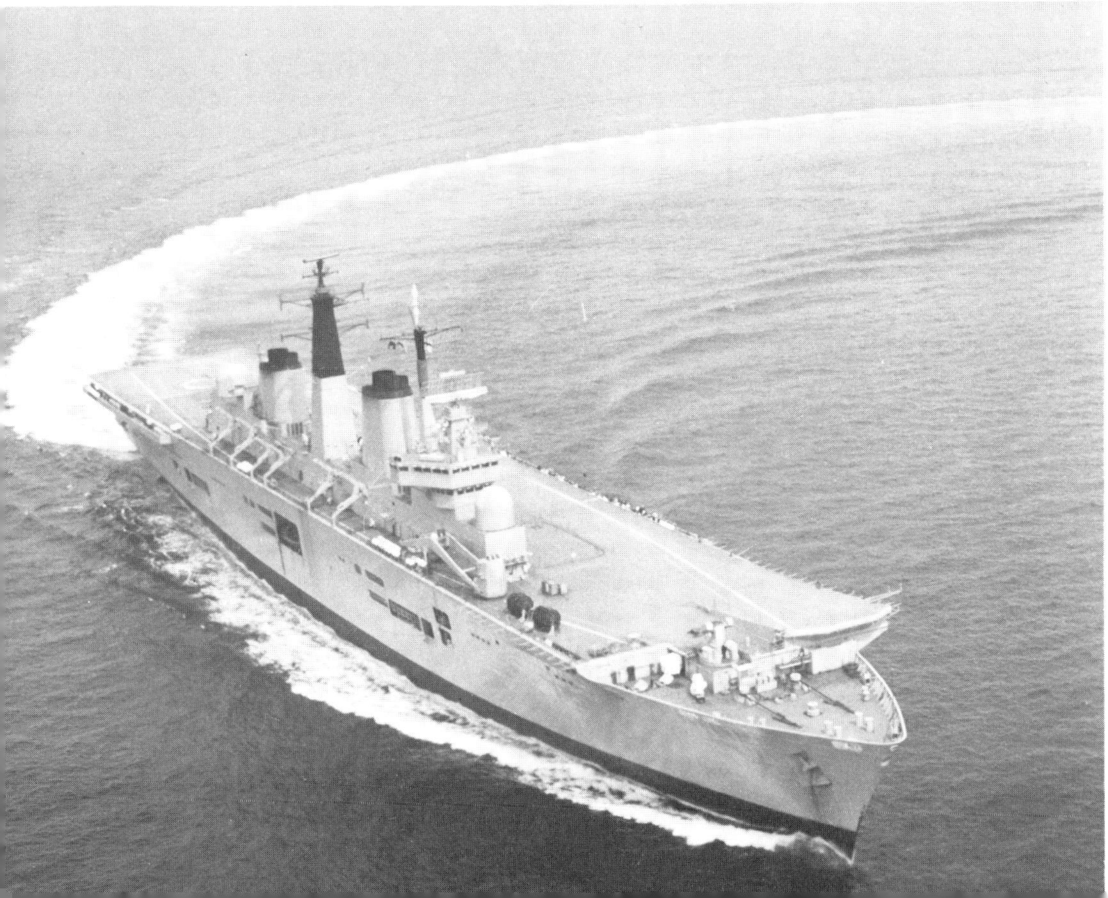

Above:
After builder's trials in 1979, *Invincible*, the first of the class, underwent Sea Trials (Naval) in the Irish Sea and although most of company had joined, including the commanding officer, the ship still wore the Red Ensign to show that Vickers retained ownership.
David Brown Gear Industries

Right:
***Illustrious* was laid down on 7 October 1976 at the Wallsend-on-Tyne yard of Swan Hunters, another part of the warship division of British Shipbuilders. This picture was taken in November 1978, just days before launching on 1 December by HRH The Princess Margaret.** *Swan Hunters*

Below right:
This view of *Illustrious* on the Wallsend slipway shows the 'A' frame supports of the propeller shafts, one of which caused the sister-ship *Invincible* so much embarrassment during 'Orient Express' when the Australian Government was reluctant to let the carrier use dry dock because of the 'nuclear weapons question'. Also clear from this view is the high freeboard of the carrier and depth of its underwater hull. *Swan Hunters*

Above:

Lying alongside the fitting out jetty at Walker Yard, Hebburn in April 1982, *Illustrious* is rapidly made ready for sea and service in the South Atlantic. The speed and efficiency of the men of Swan Hunters was a classical example of the 'Falklands Spirit' which gripped the UK at the time. The canvas covers protect the men working on the then-secret Phalanx installations. *Swan Hunters*

Right:

This aerial view shows the interesting situation of *Illustrious* being fitted out, whilst behind it, on the slipway, the first sections of *Ark Royal* are being laid. As the class has progressed, experience gained from the sea trials in *Invincible* has led to refinements and, in some cases, considerable alterations to both *Illustrious* and *Ark Royal*. *Swan Hunters*

Below right:

The massive propellers of the 'Invincible' class are driven by gas turbine engines through this large gearbox. Because the propellers do not have reverse pitch, it is the shaft which must go in reverse when 'astern' is called for from the bridge. David Brown Engineering produced the gearboxes and this picture shows the main casting being fitted with covers in November 1972. *David Brown Gear Industries*

Above:

The first reversing gearbox for *Invincible* under construction at Huddersfield, Yorkshire. *DBGI*

Above right:

The main gear unit completed with its covers attached and ready for shipping to the shipbuilders at Barrow-in-Furness. The ships have been designed to make the fitting and refitting of such major components a relatively simple task. *DBGI*

Below:

A slightly amended design for the transmission units of *Ark Royal* has been used as illustrated here. The two technicians from the manufacturers give an idea of scale. *DBGI*

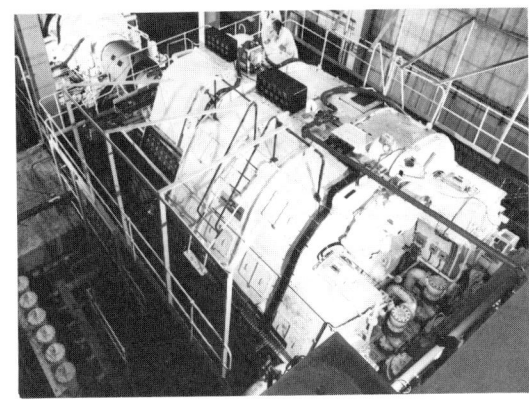

Left:
The gear units were put under test conditions with the Rolls-Royce gas turbine engines attached on shore before being fitted into the hull of each ship. This is the first unit for *Invincible* under test at the Rolls-Royce facility. *DBGI*

Below:
Swan Hunters knows the third unit of the 'Invincible' class as Ship 117. It is perhaps known better as *Ark Royal* and is seen here in April 1982 alongside the fitting out berth at Wallsend. *Swan Hunters*

Right:
Ark Royal was laid down in December 1979 and by May 1980 was beginning to take shape. *Swan Hunters*

Below:
Ark Royal in January 1981 — note the lifts and beginnings of a superstructure. *Swan Hunters*

3 Layout and Design

The light aircraft carriers of the 'Invincible' class have had added to them both roles and capabilities in the 10 to 15 years of their inception period. During that period they have gone from dual-role command and ASW ships, through the through-deck cruiser phase to that of competent light carriers (CVSs).

The class' roles can be split into two distinct types:

Primary Roles

1 The ships of the class are the modern ASW task group leaders of the NATO anti-submarine groups which need to keep the sea lanes across the North Atlantic ocean for resupply of Europe's Central Front in the event of war, and to protect the NATO Carrier Strike Fleet, centred around USN aircraft carriers, from submarine attack. In this role they were originally designed to take a maximum of 20 Sea King helicopters in overload situations, but later the design was changed to include a mix of Sea Harrier FRS1 fighters for area air defence.

2 The ships are capable of providing their own area air defence of task groups and convoys, using both the Sea Dart GWS30 missile system with associated radar guidance, and the five/six Sea Harrier aircraft in their combat air patrol (CAP) role.

3 Command, control and communications, or C³ in the NATO parlance, is the third of the primary roles. The communications facilities and the well-designed operations room can provide facilities for both naval and land force commanders, simultaneously if ncessary. These facilities have been demonstrated by *Invincible* during the 'Caribtrain' deployment to the West Indies and during Exercise 'Rough Diamond', both in 1983.

Secondary Roles

1 The ability to load troops or Royal Marines and their equipment for rapid reaction — the so-called quick dashes — is a part of the NATO and national doctrine for conventional deterrent forces. During two exercises in 1983, *Invincible* provided transportation for elements of United Kingdom/Netherlands Amphibious Force. During the deployment of South Atlantic Task Force in April 1982 *Invincible* transported RM Commandos and their equipment to Ascension Island.

2 Strike and air support to ground troops can also be carried out using the Sea Harriers aboard in their strike role (see chapter 6) or, as in the early days of Operation 'Corporate', land-based STOVL aircraft, such as the RAF's Harrier GR3, can be accommodated for strike missions. Incidental with this secondary role is the ability of aircraft embarked in an 'Invincible' class carrier to carry out physical reconnaissance missions.

3 Anti-surface ship strike can be performed by the Sea Harriers aboard, or in the future by missile-equipped Sea King helicopters which are believed to be due for deployment in the late 1980s. The Sea Dart GSW30 does have a limited anti-ship capability, but only as a weapon of last resort in terms of self-defence.

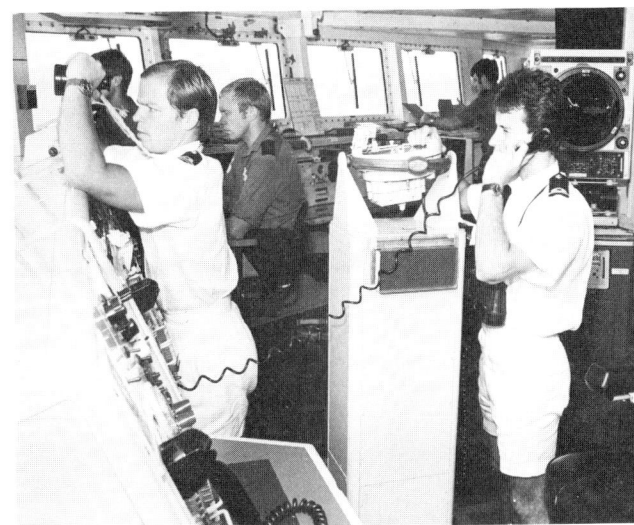

Right:
The ships' bridges are modern and spacious, with the helm situated adjacent to the officer-of-the-watch position. In the background are the communications positions (for tactical work) and the navigation radar display. *HMS Invincible*

There are obviously only a limited number of ships available for the roles mentioned above and it is not thought that all could be performed in times of tension and/or war. The decisions for the selection of roles would either rest with the Commander-in-Chief Fleet (who is also one of the three main NATO commanders, Commander-in-Chief Channel) or with the Supreme Allied Commander Atlantic (SACLANT), an American admiral based at Norfolk, Virginia.

Flagship facilities can be provided to Flag Officer First (or Second) Flotilla for task group operations, Flag Officer Third Flotilla for ASW Group operations, or Commodore Amphibious Warfare for amphibious operations, in which the ships could act as troop transports, heli-pads for Commando-support helicopters, and communications centres.

The essential design features of the 'Invincible' class are naturally dictated by the ships' roles and the present state-of-the-art in terms of naval architecture and design. The class has been described as 'space dominated' and although the displacement of around 20,000 tons seems small, the ships have an incredible amount of facilities built into the hull and superstructure. Most notable in the design as compared to previous cruiser and light aircraft designs are the engine compartments, crew facilities and accommodation, the amount of electronic and sensor equipment carried, the through hangar and the workshops. The class reflects these items, which

are found in other European and American designs, but are unique to Britain because the RN has only designed and built this one class of capital warship since the end of World War 2. Although some observers have likened the ships' high freeboard to the maintenance carrier *Unicorn* of Korean War fame, they are nevertheless impressed by the small draught. The long and thin superstructure which forms the traditional carrier 'island' did apparently cause design problems, especially in connection with the internal arrangement of sensors, access to the body of the ship (particularly the through passage deck) and the flight deck width. This latter item is particularly noticeable when an aircraft is spotted abeam the island and helicopters are operating. In fact, some

Below:
***Invincible*, and its sister-ships, incorporate various important design features. Most noticeable to those stepping aboard for the first time are the two gallery decks — Nos 2 and 5 — which allow complete access. In addition, the hangar deck layout is convenient for the workshops.** *RN*

Right:
The island superstructure contains the navigation bridge (top) and an admiral's observation platform. Normally, in the event of action, the operations room would be the key location. Note also in this picture *Invincible's* heavy-lift crane and the provision of life rafts. *Author*

Below right:
The clean freeboard of the class is a key design feature with specially situated entry ports on No 4 deck. *Author*

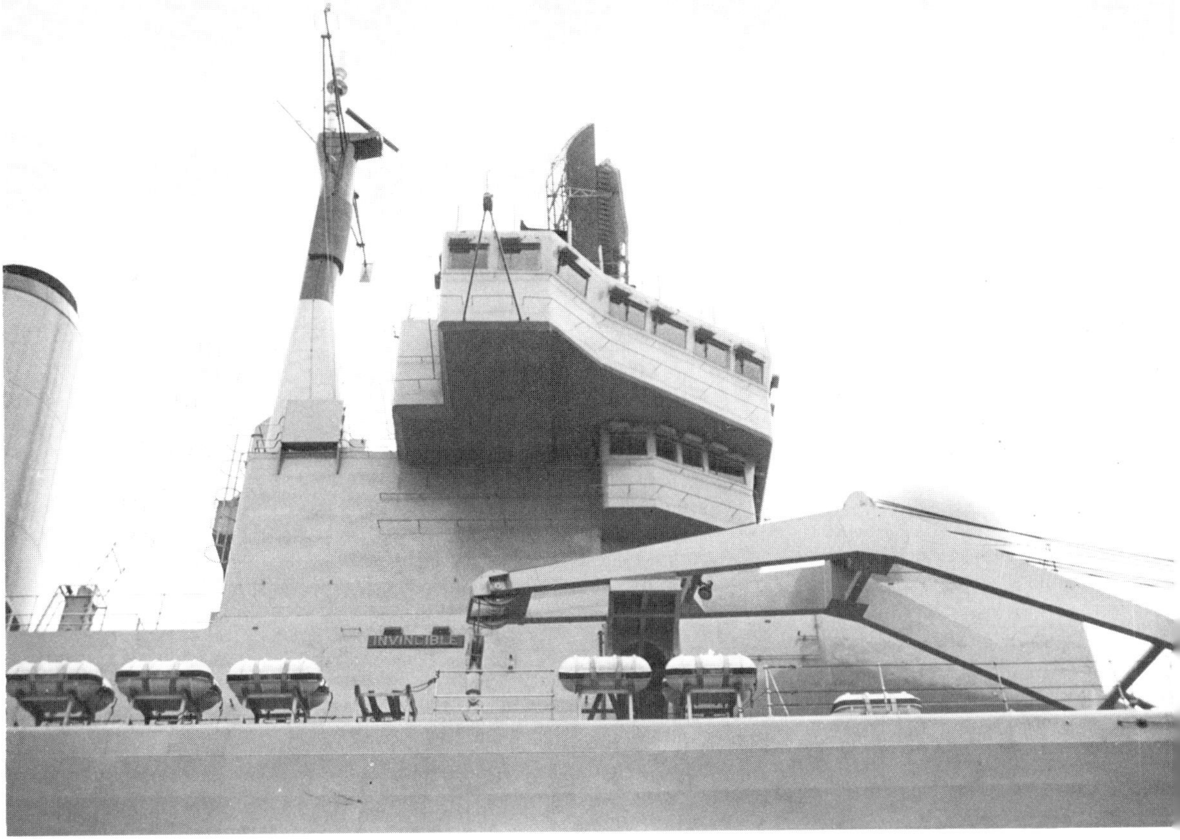
INVINCIBLE

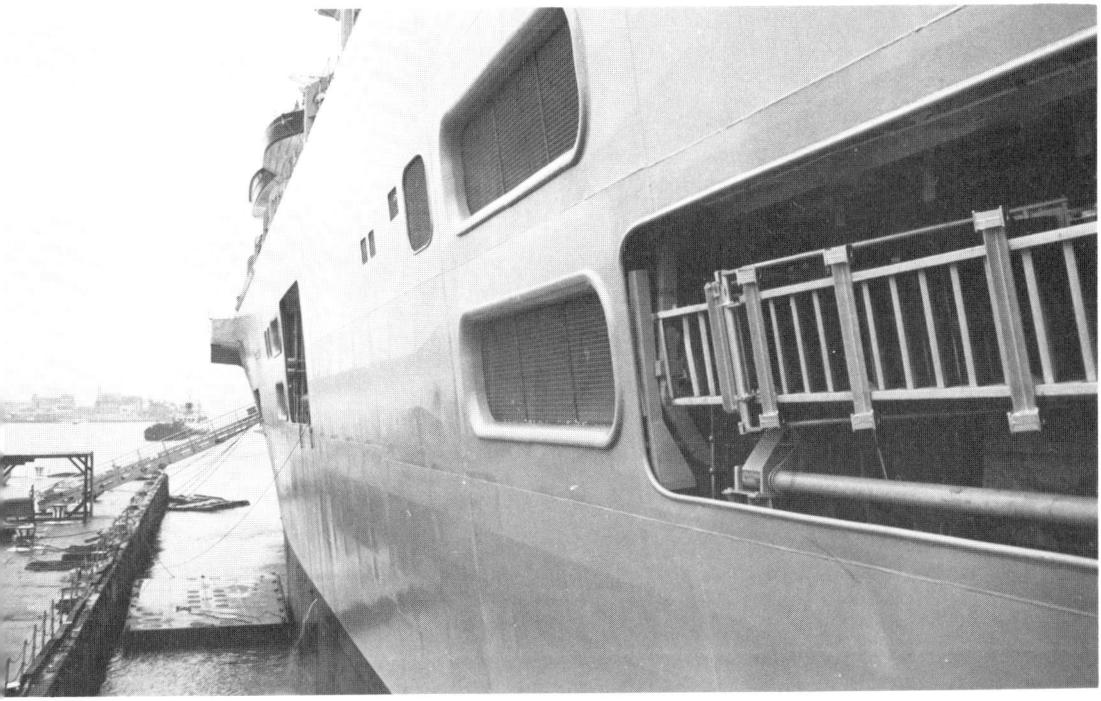

of the large types cannot fly from certain of the flight deck spots because the droop of their main rotors would strike any aircraft parked adjacent.

The aim of the hull design was to produce a good seakeeping ship and yet give little resistance to normal passage through the water. As the class would be expected to operate in the North Atlantic in the event of war, good seakeeping, especially when working aircraft, would be vital. In the South Atlantic, where the wave type is slightly different to the northern swell, the ship's company of *Invincible* reports that conditions were not particularly bad. The hull form is actually asymmetrical because the off-centre weight of the starboard side 'island' would cause imbalance if not sufficiently redressed. Twin rudders were provided to give the responsive steering again needed for an aircraft carrier, although the 'Invincible' class does not necessarily require to turn into wind for Sea Harrier or Sea King operations. The ships are fitted with two pairs of non-retractable stabilisers to keep the ship steady in most sea conditions, thus allowing the aircraft to be operating and the weapon systems to bear. The Naval Staff Requirement dictated that no more than 30% of the rolling movement should exceed ± 5deg in sea state 7, at the cruising speed of 18kt (33km/hr).

Although featuring an open fo'c'sle , the class does not have many hull-side openings and is therefore (and as a result of the hull form) a relatively dry ship. The hull and superstructure construction are, in any case, fully-welded mild steel and B-quality plating. The superstructure was wind-tunnel tested to prove that the patterns of exhaust gas discharge would not affect the flight deck operations. In addition, the effect of the superstructure to the normal wind patterns over the flight deck, especially important for helicopter operations, was also considered during this work.

Internally, the ships can be closed down for extended periods to allow for operations within areas of nuclear, biological or chemical fall-out, the carriers having filter systems and fan-driven air trunkings. For damage control purposes, the ships can be divided into separate citadels.

The ships are complex and at times confusing to the newly joined member of the ship's company, there being over 800 separate compartments in *Invincible*, and probably more in the later two ships. A warship has been described as an engineer's compromise because it must be home of its ship's company — in this case, about 1,000 men — and yet be an effective weapon of war. The Falklands conflict has proved that the class can operate successfully in the latter role and have proved happy ships in peacetime operations.

Generally, the accommodation is fore and aft, with officers and fleet chief petty officers (FCPOs) in individual cabins, while chief petty officers (CPOs) share four to a cabin, petty officers (POs) are in groups of six, and there are never more than 18 junior ratings in individual messes. Even compared to *Hermes* or a 'County' class destroyer, this provides high standards which can only be good for morale. In overload conditions, such as some of the operations in Operation 'Corporate', *Invincible* did become rather overcrowded with officers sharing cabins, or even sleeping on camp beds in passageways. During Commando operations, the embarked forces would expect to be accommodated in the hangar.

Food is important to every sailor. There is a centralised main galley which serves the three separate dining halls for FCPOs and CPOs, POs, and Junior rates. A further, smaller galley serves the wardroom and staff officer/flag officer accommodation aft.

Although the 'Invincible' class took several years to design, caused mainly through political mishandling, the design was co-ordinated and integrated. Nevertheless, the fitting in of the machinery spaces and control rooms did cause some early problems, mainly because the philosophy of gas turbine propulsion was new to many engineers. The machinery aspects of the design are discussed in the next chapter.

The most dominant feature of the ship's design is the main flight deck which effectively consists of a short take-off runway some 550ft (167.6m) by 40ft (12.2m), angled half a degree to port to clear the Sea Dart launcher and its blast screen. In addition there were six (later changed to eight) helicopter spots along the length of the flight deck. Sea Harriers are run up aft, in four positions clustered around the stern end of the 'runway'.

Along the catwalks around the flight deck are facilities for firefighting, aircraft fuelling and defuelling, ground electrical supplies, and aft, the MEL MADGE Harrier landing aid.

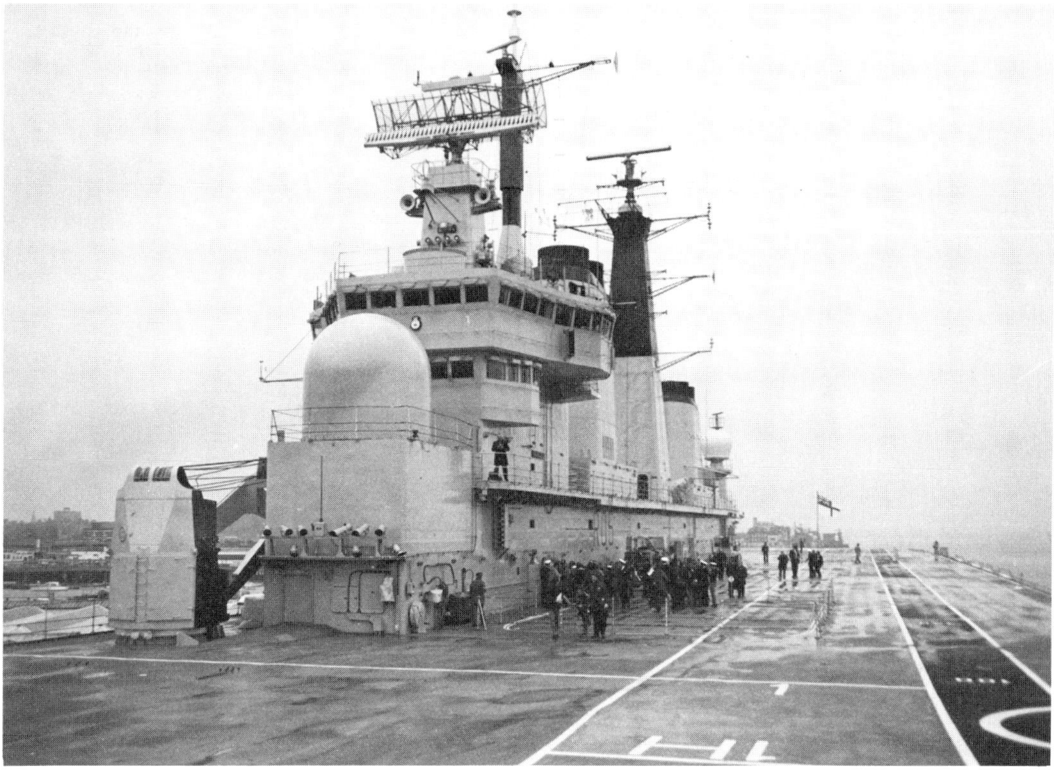

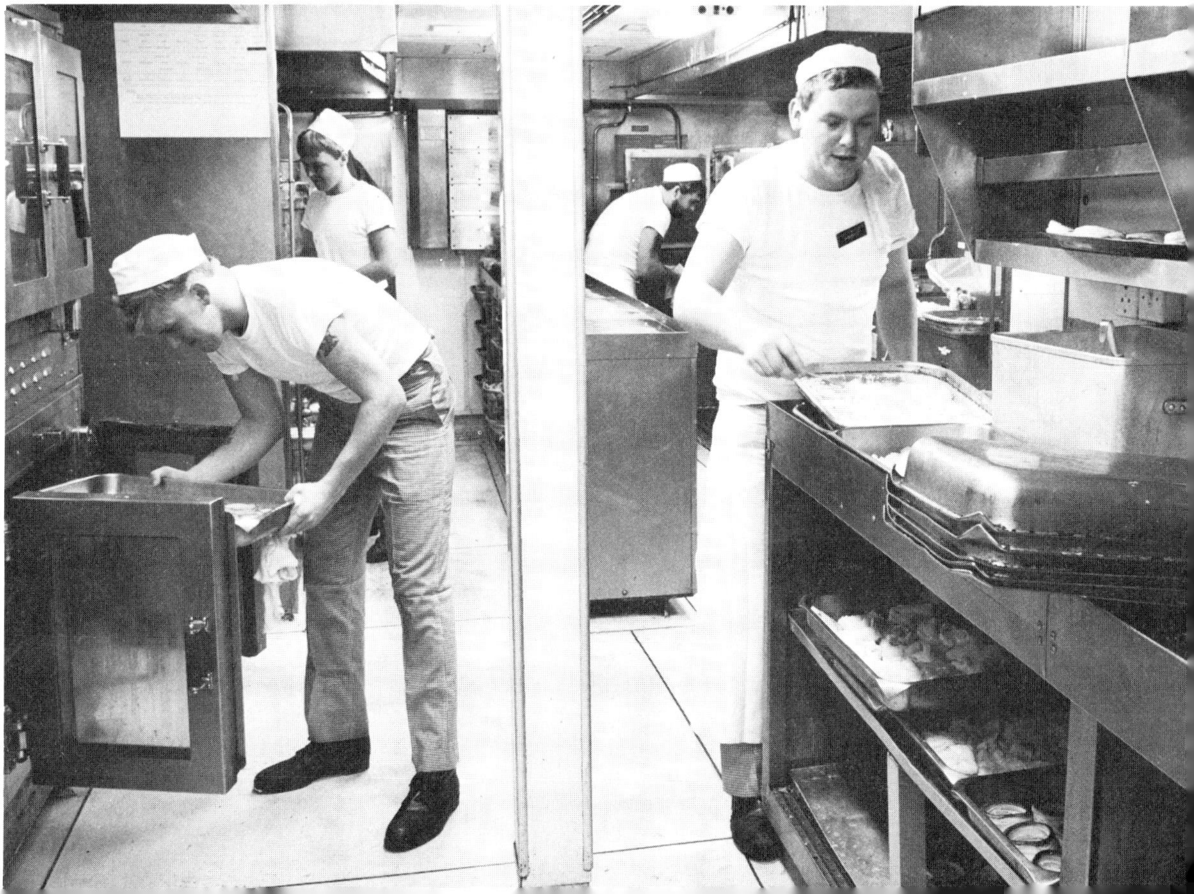

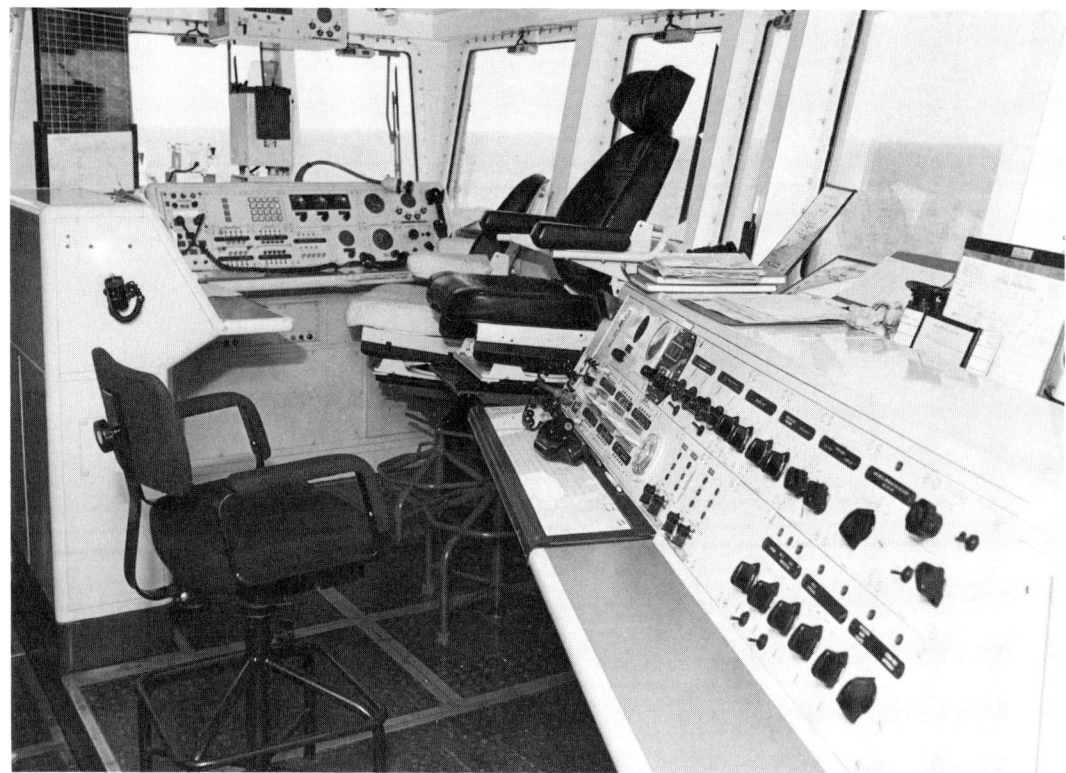

Above:
During flying operations, the Flying Control — or 'Flyco' — is continuously manned by an officer, Lieutenant Commander (Flying), and his rating logger. There is an additional position for the Commander (Air) — the high chair — who spends the bulk of his time on 'Flyco'. The position has direct access to the navigation bridge and a good outlook down the deck. *HMS Invincible*

Operations on the flight deck are controlled by the Lieutenant Commander (Flying) - 'Little F' in FAA parlance — and the Flight Deck Officer ('Fido'). 'Little F' has a seat in the Flying Control Centre ('Flyco') from where he has communications facilities to direct aircraft in the launch, recovery and circuit modes, on the flight deck and in the air generally, but only at short range. The flight deck personnel, including the Flight Deck Officer, communicate with each other and 'Flyco' using a closed circuit radio link. Additional communication lines are used to other departments in the ship, including the Hangar Control Officer. Flyco is situated on the level but aft of the bridge, below which there is an admiral's bridge. Other compartments in the island include accommodation for the aircraft handling parties and fire fighters, flight deck stores, briefing rooms and weapons preparation.

The hangar deck is the main compartment of the internal layout and its size and shape has dominated the design of the 'Invincible' class. Immediately off the hangar deck is the suite of workshop facilities with individual service compartments for aircraft mechanical engineering, engine repair, component repairs — including ejection seats, sonar sets, instrument maintenance, air ordnance (missiles and torpedoes), batteries and sensors (radar) — and even a workshop for parachute maintenance. These compartments are independent of the fire curtain system which divides the hangar into three. In normal conditions with the present air group, 17 aircraft can be accommodated, but more would require parking space on the flight deck.

The elevators for transporting aircraft to the flight deck from the hangar deck, called 'lifts' in the RN, are non-traditional in design. In older carriers, aircraft lifts have been carried on special multiple chains on two sides, driven by electric motors. This has not been the best solution but the opportunity had not arisen before for redesign. The 'Invincible' class building programme has allowed MacTaggart Scott, the world-famous ship machinery manufacturers, to propose a less weighty hydraulically-operated lift. This lift, of which two are found per ship, has not only the advantage of doing away with the heavy counter-balance weights but also allows access

on more than just two sides as with previous designs found on traditional carriers like *Hermes* and *Bulwark*. Damage control experts will also tell you that lifts so constructed with chains are more liable to be put out of action by direct, or indirect, enemy action, including blast damage. An aircraft carrier without functioning lifts is virtually useless in its primary role. The new lifts are made of steel with Y-shaped prefabricated box construction sections, the upper ends of which support the lift platform (which also forms part of the flight deck structure), and the lower ends are attached to travelling rollers on a special track. The tracks help to distribute the load into the ship's structure and steady the motion of the platform. On one side of the 'Invincible' class lifts there are guides to restrain unusual movement, but the penalty of this is to allow loading and unloading from only three sides. Nevertheless, in a small hangar, the lifts must be positioned so as not to cause an obstruction. The hydraulic power comes from electric motors but, as a battle damage pre-

caution the system is inter-connected with other ship's systems.

MacTaggart Scott was involved in numerous tests onshore and is reported to have carried out over 20,000 trial runs with a mock-up lift. The company has stated that the lift actually weighs less than half a traditional lift of the same dimensions.

The hangar deck also boasts hangar bulkhead services as found on the flight deck and mentioned above, as well as a travelling gantry for engineering use. This allows engine changes in the Sea Harriers and the Sea Kings which would otherwise be difficult in shipborne conditions. Helicopter rotor heads, gearboxes and other transmission components can be worked on or replaced with this machinery. Additionally, flight deck equipment comes in two varieties — the fixed gear on the deck and the movable equipment which can be used in the hangar, as well as on the flight deck.

The main feature on the flight deck, apart from the Vulcan Phalanx gun (see chapter 6) is the fixed crane unit, positioned forward of the island and used for recovering ditched aircraft and for moving aircraft from the flight deck to a quayside or aircraft movement lighter. The movable equipment includes a number of AWD (all-wheel-drive) tractors, a JCB telescopic lift, a number of Hyster industrial forklifts and a Coles mobile crane. This latter vehicle was specially built and rushed to *Illustrious* when the ship was being hurriedly completed for sea duty.

All-Wheel-Drive Tractor

This vehicle was designed for handling aircraft on the carrier flight deck, with particular reference to the spotting of aircraft. The overall dimensions have been kept to an absolute minimum to aid mobility, and the towing gear designed to aid manoeuvrability during the positioning of both fixed-wing and rotary-wing aircraft.

Length: 12ft
Width: 6.67ft
Ground Clearance: 0.67ft
Weight: 5.75 tons
Turning Circle: 24ft

The AWD tractor is powered by a 122bhp Perkins direct injection diesel engine, giving speeds of 4mph (low gear) and 10mph (high).

JCB Telescopic Lift

This is a front wheel drive, rear wheel steer machine which is extremely versatile and therefore fits into the carrier inventory well. It has capabilities well beyond the scope of conventional fork lifts and is an important tool in the maintenance and cleaning of aircraft, fire-fighting and rescue (where rescue workers can be lifted directly to the cockpit of a Sea Harrier to extract the pilot), and stores handling. The JCB520 can lift 2.25 tonnes to 21ft.

Hyster Industrial Forklift

The Hyster forklift is designed for speedy and efficient handling of relatively heavy stores on board ship. The machine is used in the hangar of 'Invincible' class carriers, as well as on the flight deck for materials handling.

Coles Speedcrane

An important addition to the range of flight deck vehicles carried in a modern carrier, the Coles Speedcrane is used for lifting the heaviest loads on the flight deck. It can also be used for engine and transmission changes, and other engineering tasks. In the event of a crash on deck, the Coles would be used for removing the wreckage or damaged aircraft. The crane is one-man operated, fully mobile and slewing, with power supplied by a Ford 2713E diesel engine which develops 100bhp at 2,500rpm.

Construction and Design of the Ski-Jump

One of the most impressive features of the 'Invincible' class' design is the slightly angled ski-jump ramp which was fitted to *Invincible* during building, but which was designed into the construction programmes for *Illustrious* and *Ark Royal*. The aim of the system is purely to enable fully-loaded Sea Harrier STOVL fighters to be launched in almost any wind conditions. For this purpose both *Invincible* and *Illustrious* have 7deg ramps, whilst *Ark Royal* has been fitted with a 15deg system. Although wartime carriers were fitted with wooden ramps to assist the performance of heavily-laden torpedo bombers, such as the Fairey Barracuda, the ski-jump idea is novel

Below left:
Looking forward from the amidships area, the ski-jump ramp does not look particularly significant, but its effect on the performance of the Sea Harrier is critical to the operating of these aircraft. *Author*

Above:
Initially trials with various angles for the ski-jump were carried out at Royal Aircraft Establishment Bedford. Illustrated is a Sea Harrier FRS1 from 899 Squadron, the headquarters unit normally based at Yeovilton, but which embarked in both *Invincible* and *Hermes* for Operation 'Corporate'. *801 Squadron/A. P. Amesbury*

Below:
The ski-jump in use. Lt Ian Watson launches into space during flying operations in the Tasman Sea when 801 Squadron was embarked in *Invincible*. The ship on the horizon is RFA *Regent*, one of the group's support ships in the Far East. *Author*

and is directly attributable to one man — former Lt Cdr Doug Taylor MPhil, MBE, who retired in 1979 and now works for the Marconi Avionics Future Systems Group.

The ski-jump is technically straightforward and of low cost to construct, and gives an improvement in performance in a critical area of flight. It only took two years to develop fully, although the idea had been mooted by Doug Taylor almost eight years before. During the Indonesian Confrontation, Taylor was Flight Deck Engineer in *Victorious*, which he describes as 'a slick ship', especially in the flight deck department. During a passage through potentially hostile waters, en route Perth to Singapore with a Sea Vixen fighter ready to launch on each catapult, action stations sounded and the aircraft were prepared for immediate launch. Unfortunately, both catapults jammed due to the intense heat of the sun expanding the steel runners. This problem lasted only for a few minutes and then the steam catapults — arguably the ship's main armament — freed and the aircraft were successfully launched. This incident was to play a major part in the development of the Sea Harrier concept in connection with the 'Invincible' class aircraft carriers a decade later.

In his spare-time, Doug Taylor toyed with the idea of fitting Sea Harriers on to 'Leander' class frigates and considered V/STOL aviation generally, especially because it was becoming apparent that the Sea Harrier with a full fuel and war load would need about 980ft (299m) to take off from a flight deck, even with wind over the deck. A vertical launch catapult was discarded, but an inclined catapult would give better performance with an optimum angle of 60deg; this amount is just not practicable and it was found that 20deg would be sufficient, although less 'would do'. There was therefore a solution to the launching of fully laden Sea Harriers from the new generation of aircraft carriers, then still called through-deck cruisers. Various systems, including an inflatable catapult, were also discussed, but the political climate did not allow for further work directly within the RN. It seemed likely that any jets embarked in the new 'cruisers' would need to take a short run down a completely clear deck to become airborne. The 'Runway in the Sky' was first seriously considered at Christmas 1968 and Taylor announced that his calculation had determined that a ski-jump would provide a cheap, improved-safety performance-enhancing launching system, which although it meant doubling the take off distance of the Harrier, would ensure take off. The ski-jump penalty is that the jet does not fly for several hundred yards after leaving the ship, but follows a semi-ballistic trajectory.

At this time the Sea Harrier concept had not been officially sponsored and the 'Invincible' class' future was still in doubt. Various problems and faults were raised with Taylor, but he had answers for them:

What if a tyre burst?
The Sea Harrier would just make a black streak down the flight deck.
What if the engine fails?
Although the aircraft might be lost, the pilot would have three times the normal naval jet ejection time, even once on the 'trajectory'.
What about crosswinds?
These do not affect the operation of the ski-jump.
What if the ship pitches?
Again there is no problem for the aircraft.

The Ships' Departments

Although all three warships of the 'Invincible' class have different compartment layouts between decks, their general arrangement in terms of personnel is basically the same. For ease of operation personnel are divided into 'Departments', the head of which — referred to as a 'HOD' — is usually a Commander.

Executive Department
Under the Captain of an 'Invincible' class ship is the Commander or Executive Officer, who is responsible to the Captain for the smooth and efficient running of the ship. In addition he is President of the Wardroom Mess and runs the Ship's Welfare Committee. He is assisted by the First Lieutenant, who is also the ship's Security Officer; between them, these two officers administer all discipline not serious enough to be taken to the Captain's table. Also within the Executive Department is the Navigating Officer, who 'drives' the ship; the Senior Warfare Officer (SWO), who 'fights' the ship; and the Torpedo Anti-Submarine Officer (TAS), also known in more modern naval parlance as the Advanced Warfare Officer (Underwater) — AWO(U). Because of the size and importance of the class, the ships carry a specialist Signal Communication Officer (SCO) who deals with all communications from the tactical level to the transmissions via satellite. He is also concerned with the problems of RadHaz (radiation hazards) from the ship's radios, and electronic warfare (EW) policy. Other senior members of this department include the NBCD (Nuclear, Biological and Chemical Defence) Officer and the Principal Warfare Officer (Air) —

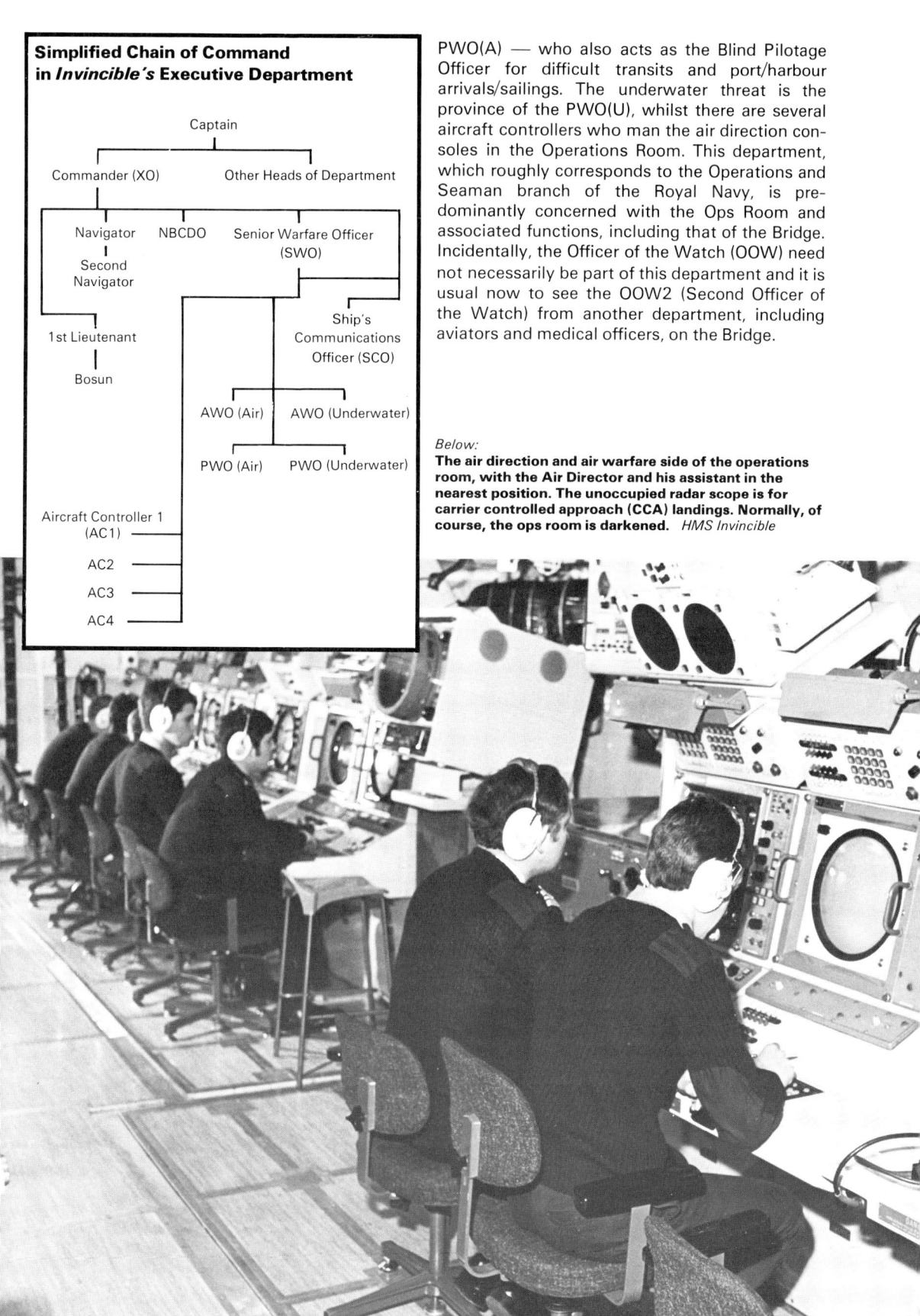

Simplified Chain of Command in *Invincible's* Executive Department

```
                        Captain
          ┌───────────────┼───────────────┐
    Commander (XO)                 Other Heads of Department
  ┌────────┼──────────────────────┐
Navigator  NBCDO          Senior Warfare Officer
   │                            (SWO)
Second                    ┌────────────┴──────┐
Navigator                              Ship's
   │                               Communications
1st Lieutenant                     Officer (SCO)
   │                    ┌──────────────┐
Bosun               AWO (Air)    AWO (Underwater)
                    ┌─────┴────────────┐
                 PWO (Air)    PWO (Underwater)

Aircraft Controller 1
      (AC1) ─────────
       AC2 ─────────
       AC3 ─────────
       AC4 ─────────
```

PWO(A) — who also acts as the Blind Pilotage Officer for difficult transits and port/harbour arrivals/sailings. The underwater threat is the province of the PWO(U), whilst there are several aircraft controllers who man the air direction consoles in the Operations Room. This department, which roughly corresponds to the Operations and Seaman branch of the Royal Navy, is predominantly concerned with the Ops Room and associated functions, including that of the Bridge. Incidentally, the Officer of the Watch (OOW) need not necessarily be part of this department and it is usual now to see the OOW2 (Second Officer of the Watch) from another department, including aviators and medical officers, on the Bridge.

Below:
The air direction and air warfare side of the operations room, with the Air Director and his assistant in the nearest position. The unoccupied radar scope is for carrier controlled approach (CCA) landings. Normally, of course, the ops room is darkened. *HMS Invincible*

There are eight Paxman diesel generators to supply the various requirements for the class's different compartments. Usually screened off to prevent excessive noise, one of *Invincible's* generators is being overhauled by MEM Alan Stewart and MEM 'Smokey' Coles of *Invincible's* Marine Engineering Department. *HMS Invincible*

Marine Engineering Department

The head of this department is the Marine Engineering Officer, also known as Cdr(E), and he is responsible for the ship's propulsion, hull, flight deck equipment, domestic machinery, electric generation and distribution, and hydraulics. During the last few years the overlap with certain areas of the Weapon Engineering Department (see below) has been resolved, leaving the ME Department with its responsibility of keeping the ship running. The advent of the gas turbine main propulsion system has allowed a decrease in all ranks and rates of this department.

The ME Department controls the functions of the ships in the class from the Ship Control Centre (SCC), formerly known as the Machinery Control Room (MCR) in other classes, which is situated adjacent to the Damage Control Centre (HQ1).

Below:
One of the landing systems used to aid the recovery of Sea Harrier aircraft is the Deck Approach Projector Sight — known, not surprisingly, as DAPS — situated adjacent to the 'Flyco' position. The vertical series of lights are aligned with the horizontal bars by the recovering pilot to obtain the correct angle of approach. This is a modified version of the project sight found near the arrester wires in conventional aircraft carriers in the past. *Author*

Weapon Engineering Department

Today, the Weapon Engineering branch of the RN has three main functions: the care of sensors (mainly radar), weapons (mainly guided missiles), and communications (including satellite systems). Onboard an 'Invincible' class aircraft carrier, the Department is headed by the WEO (Weapon Engineer Officer), who has a deputy (DWEO), as well as three specialist engineers for the three facets of the work. WE personnel are nicknamed 'greenies'. The Type 1022 radar and the Operations Room set-up are this department's two greatest responsibilities, followed by the ships' primary weapon system, the Sea Dart guided missile and its associated Type 909 radars.

Supply and Secretariat Department

Every piece of equipment aboard the ship is borne on the ledgers of the Supply and Secretariat Department. Beside the Captain's office (under the Captain's secretary), the Department, headed by the Supply Officer (SO), deal with three particular sub-specialisations: Catering, which includes the provision of victuals to the Wardroom; Chiefs', POs' and Junior Rates' messes; Cash, which deals with all pay and money matters; and Stores.

Air Department

The Air Department, headed by Commander (Air), or 'Wings', is responsible for air operations from the ship. The tactical and local control is provided by Lieutenant Commander Flying, or 'Little F', from his control position overlooking the flight deck, on the same level as the bridge. To aid him the ship has two Helicopter Approach Position

Indicators (HAPIs) and a modern carrier landing aid, known as DAPS (Deck Approach Projector Sight) — which was developed from the former mirror sight of the angled-deck carriers — as well as the proficient flight deck team, under the Flight Deck Officer (FDO). The Department is also responsible for Flight Safety and photography including the developing, printing and analysis of intelligence and training material taken by the Sea Harrier's camera equipment or the portable equipment carried in the Sea King. There are various types of aircraft handler, whose specialist roles are shown by their coloured waistcoats — for example, green is for electricians, yellow for directors, red for armourers, etc. Air traffic control, either from Flyco, or more usually from the Ops Room, comes within the gambit of the Air Department, as do air operations.

Air Engineering Department

The raison d'etre of an aircraft carrier is to fly its aircraft, and although each embarked air squadron will have its own engineering personnel embarked also, the ships provide back-up services in the form of specialist workshops for the inspection and repair of air weapons (including the AIM-9L Sidewinder AAM), radios and avionics (including the Sea Harrier's Blue Fox radar and the Sea King's Plessey dipping sonar), flight deck equipment, electrical air maintenance (including basic flight instruments, controls systems and inertial navigation equipment) and survival equipment. The modern aircrew has special survival suits, known popularly as 'goon bags', life preservers (with built-in radio and signalling equipment), dinghies, and helmets (known as 'bonedomes'). Sea Harrier pilots also have oxygen equipment and parachutes.

The head of Air Engineering Department is the AEO (Air Engineering Officer), who is supported by the Specialist Air Weapons Engineering Officer, the Air Control Room Officer (ACRO), an Explosives Safety Officer and a Mechanical Air Engineer (MAEO). In addition there are specialist technicians of CPO, PO and leading rates.

The Education and METOC Department

The term METOC is derived from METeorology and OCeanography, two sciences which are combined in the modern Royal Navy. On an aircraft carrier it is especially important to know exactly what the weather is doing, because the safe and effective flying of aircraft depends on the weather

Left:
The ship is equipped with some very modern and sophisticated aviation engineering equipment and, naturally, has the craftsmen to operate it.
HMS Invincible

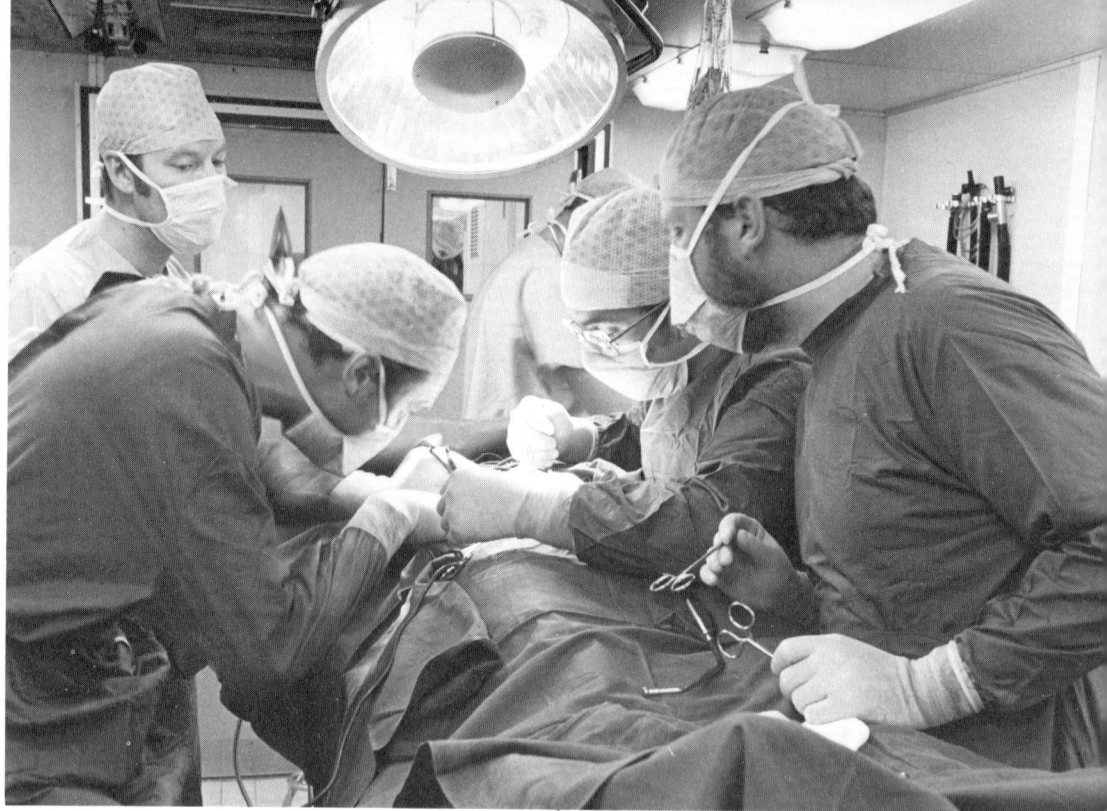

The ships, being generally expected to head task groups or forces, are well equipped with medical supplies, an operating theatre, a large sick bay and have a surgeon commander as head of the Medical Department. There is, in addition, a dental surgery. *HMS Invincible*

and the sea state. Should the sea become very rough, then the flight deck is deemed to be 'out of limits' and cannot safely, in peacetime, be used for flying. Although the figures have yet to be released, it is thought that *Invincible* performed well in the South Atlantic in 1982, despite the rough conditions; it is important to consider this aspect when assessing whether a conventional, fixed-wing aircraft carrier would have done 'better' with a larger air group. Oceanography is important to the anti-submarine and underwater warfare roles of the class, where such items as sound propagation and sea temperature are factors which can affect the range and performance of the sonar equipment, both passive and active.

A large ship in the modern RN has a real need for an effective education programme, not only in terms of civilian 'O' and 'A' level qualifications, but also dealing with in-Service examinations. The educationalists combine their role with that of the METOC, but also deal with public and press relations, classified books and papers, entertain-ments and visits. The head of this department is the Senior Instructor Officer (SIO).

Medical Department

Again, because of their size, the 'Invincible' class carriers are fully fitted with sick bay, operating theatre and ward facilities. The PMO (Principal Medical Officer) is supported by a Medical Officer (a Surgeon Lieutenant Commander), a CPO Medical Assistant, with a number of PO and junior rates. The weapons lift from the flight deck can be used to bring wounded or injured personnel to the sick bay level, thus alleviating the PMO's workload in an emergency. Throughout the ships there are first aid posts and emergency stretcher positions.

The Medical Department also deals with Health and Safety at Work and radiation safety.

Dental Department

The well-equipped dental surgery is staffed by the Senior Dental Surgeon (of Commander rank) with one senior rate assistant.

Chaplaincy

The ships each carry a chaplain, who looks after the spiritual and welfare needs of the ships' companies. In addition, his role includes resettlement of those going out of Service, and families' liaison work.

The Sea King Squadron

The embarked Sea King anti-submarine warfare squadron is also tasked with providing search and rescue, helicopter delivery service (HDS), and surface search facilities (with radar or visually). The squadron can embark up to 10 Sea Kings with full maintenance facilities for all but 'deep' maintenance. The squadron is under the command of a senior Lieutenant Commander, with the assistance of two further Lieutenant Commanders who act as Senior Pilot (SPLOT) and Senior Observer (SOBS). The SOBS also acts as Operations Officer, although this role could be delegated to a senior Lieutenant. Such offers provide the Qualified Helicopter Instructor (QHI) for pilot check flying and the Observer Training Officer — formerly the Jezebel (passive sonobuoy) Training Officer. The Observer is assisted in the rear cabin of the Sea King by the aircrewman, who can be a leading rate, Petty Officer, CPO or Fleet Chief. Through the Senior Aircrewman, they report to the Senior Observer. The Commanding Officer of the squadron reports, via the Commander (Air) to the Captain of the ship. RN Sea King squadrons are based at RNAS Culdrose when not embarked. The CO can be either a pilot or an observer.

The maintenance structure of the typical Sea King squadron is based on the Air Engineering Officer (AEO) who reports to the squadron CO. Under him are the Deputy AEO and the Assistant AEO. The Senior Maintenance Rating (SMR) controls the radio, weapons, mechanical and electrical ratings who keep the aircraft and their systems airworthy.

The Sea King AEW flights embarked in the class are operating initially 'under the wing' of the Sea King squadron, but it is understood that future plans will include the formation of AEW flights in the style of the Fairey Gannet AEW3 flights embarked in the fixed-wing carriers. When such flights of two or three Sea King AEW helicopters are formed, they will have their own reporting structure with CO, Senior Pilot, Senior Observer, AEO, etc.

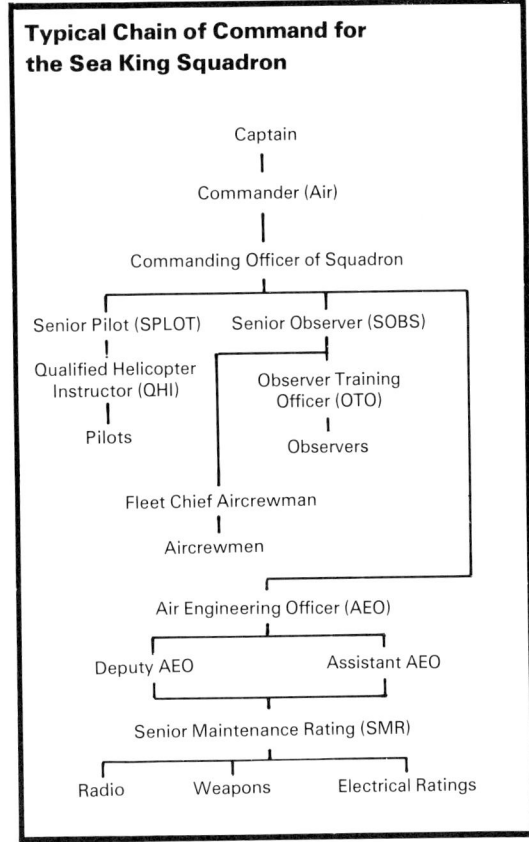

Typical Chain of Command for the Sea King Squadron

Captain
|
Commander (Air)
|
Commanding Officer of Squadron
|
Senior Pilot (SPLOT) — Senior Observer (SOBS)
|
Qualified Helicopter Instructor (QHI) — Observer Training Officer (OTO)
|
Pilots — Observers
|
Fleet Chief Aircrewman
|
Aircrewmen
|
Air Engineering Officer (AEO)
|
Deputy AEO — Assistant AEO
|
Senior Maintenance Rating (SMR)
|
Radio — Weapons — Electrical Ratings

Below:
A formation of Sea King HAS5 helicopters from 820 Squadron, embarked at the time in *Invincible*. Although embarked for the Falklands War, 820 Squadron has also operated from *Illustrious*, including the 'Ocean Safari' exercise in early summer 1983. *HMS Invincible*

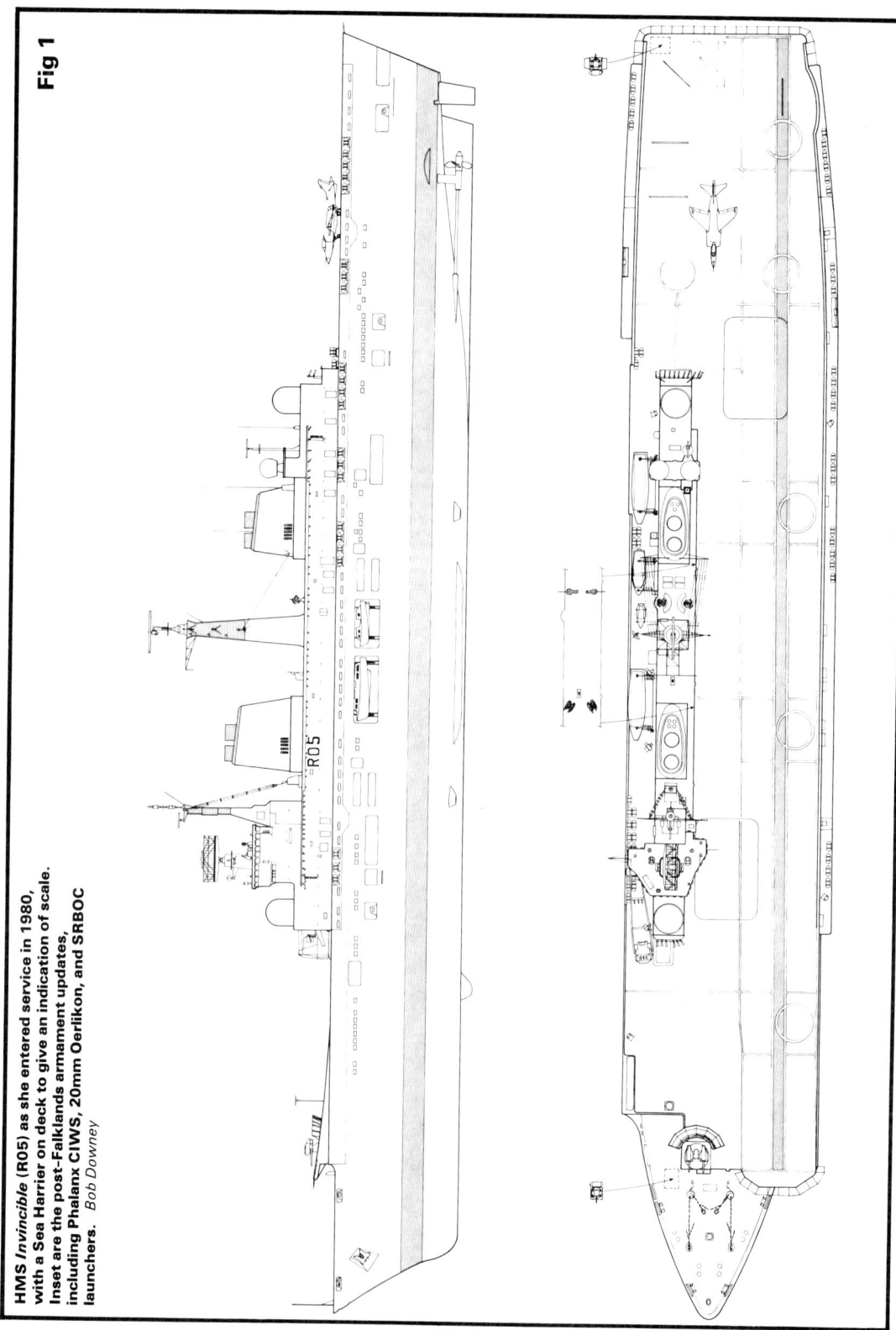

Fig 1

HMS *Invincible* (R05) as she entered service in 1980, with a Sea Harrier on deck to give an indication of scale. Inset are the post-Falklands armament updates, including Phalanx CIWS, 20mm Oerlikon, and SRBOC launchers. *Bob Downey*

Fig 2

HMS *Illustrious* (R06) as she entered service in 1982 with Phalanx CIWS and other close range defences already fitted. Note the slightly different positions for the weapons and the improved flight deck markings (although the exact details have been deleted for the sake of clarity). *Bob Downey*

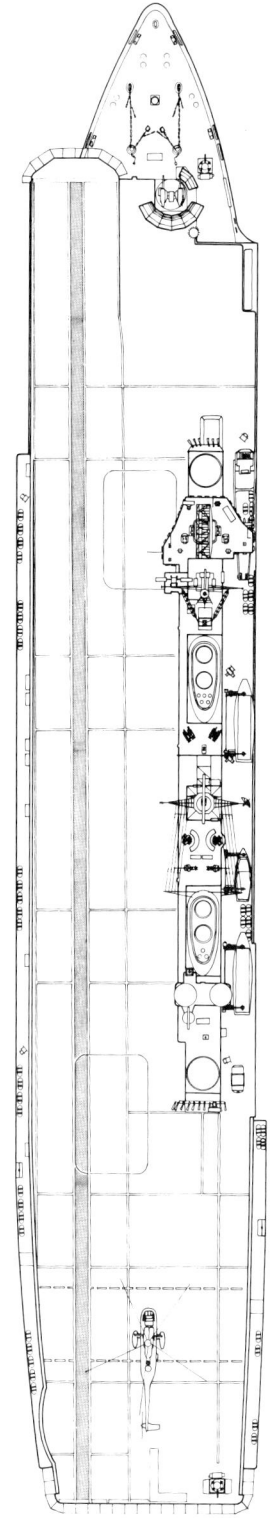

The Sea Harrier Squadron

Although not originally envisaged as being part of the 'team' aboard the 'Invincible' class, the Sea Harrier squadron has fitted into the scheme well. The normal complement is six aircraft and 10 pilots. In overall command is the Commanding Officer, with the aircrew reporting to him through the Senior Pilot, who is also the Executive Officer of the squadron. Other pilots have the responsibility for such roles as Qualified Fighter Instructor (QFI), Air Warfare Instructor (AWI) and First Lieutenant.

On the maintenance side, the Air Engineer Officer (AEO) reports to the CO, and the maintenance personnel are split into two watches under the Senior Maintenance Rating (SMR), who reports to the AEO, and the Deputy SMR, who reports to the Deputy AEO.

Below:
Immediately before the Falklands conflict, *Invincible* had exercised its amphibious forces carrying role off the coast of Norway. The Sea King helicopters of the resident air squadron (820 at the time) are also capable of lifting the vehicles, stores and equipment used by the modern Royal Marines. *RN*

Typical Chain of Command for the Sea Harrier Squadron

Captain

Commander (Air)

Commanding Officer of Squadron

Senior Pilot (XO) — Air Engineer Officer

Pilots — 1st Lieutenant — QFI — AWI

Regulating CPO — DAEO

SMR — DSMR

Maintainers

Mechanical

Radio

Weapons

Electrical

Invincible and *Illustrious* — External Differences

The major external differences between *Invincible* and *Illustrious*, following the former's return from the South Atlantic in September 1982, and the fitting of the SRBOC, 20mm GAM-BO1 and Phalanx systems, are listed below:

1 The flight deck crane has been moved farther aft in *Illustrious*, along the Alaskan Highway, giving more space to park aircraft forward of the bridge.
2 The UHF/DF aerial in *Invincible* is raised vertical by hydraulics, whereas that in *Illustrious* is a folding type, placed farther forward.
3 *Illustrious's* forward Phalanx system is nearer the bow.
4 The bridge top of *Illustrious* has less space compared to *Invincible*.
5 There is a difference in the number of life rafts along the side of the flight deck which will be evident from close study of the photograph of the two ships steaming together in the South Atlantic, on page 81.

Ark Royal's Arrangement

At the time of writing, it is reported that *Ark Royal*, although outwardly generally similar to the other two ships in the class, will have a different flight deck and internal arrangement. The upper deck seaboats have been omitted from the design, and the space along what is known as the 'Alaskan Highway' will be used to deck park Sea King helicopters. The flight deck will be extended on to the fo'c'sle necessitating a repositioning of the Sea Dart launcher and the forward Phalanx gun. The Phalanx arrangement has in any case been altered to give three guns available: either forward, amidships (on the superstructure) and aft, or forward and sponsons on each quarter. This rearrangement would give a far better angle of fire against incoming missiles and strike aircraft. It is also reported that provision has been made for the placing of vertical-launch Sea Wolf missiles for close-in defence against attacking aircraft and missiles. The Sea Wolf launchers would be placed forward of the flight deck park on the fo'c'sle. The ski-jump ramp angle has been extended to 12deg in an effort to increase the load carrying ability of the Sea Harrier FRS1 and the updated version which might come into service as early as 1986.

Finally, Ark Royal's hangar space has been rearranged and improved, aided by a rearrangement of the accommodation spaces and workshops. The extra hangar space will be required for the addition of Sea King AEW helicopters and the workshops to overhaul their Searchwater radar system. Such hangar space reorganisation will be carried out in *Invincible* and *Illustrious* when these ships come in for refit in 1985-86 and 1988-89 respectively.

Below:
Illustrious is of very similar appearance to its earlier sister-ship, although the internal layout has been amended in certain areas following experience gained in Invincible's operation. In nearly all respects, the changes are minimal. *Mike Lennon*

Above:
Ark Royal pictured lying in the River Tyne during March 1984, during the fitting-out phase of its building programme. Note the steepness of the ski-ramp and position of the deck crane. Another recognition feature is the four exhausts on the funnel tops. *Fleet Air Arm*

Below:
Although this is a good overhead view of **Ark Royal**, it is not possible to identify the positions of its CIWS armament, nor to identify any other special features. It is certain that the bulk of experience learned from operating the two previous ships in the class has been put into the internal arrangement. *Fleet Air Arm*

4 Machinery

When the final design studies had been completed in 1970, the 'Invincible' class ships were destined to become the largest warships in the world with gas turbine machinery, including the RN's largest power shaft, propellers and electrical power generation plant. The ships are rated as COGAG warships, that is to say that they have COmbined Gas And Gas propulsion machinery — four Rolls-Royce Olympus TM3B Marine gas turbines. In addition, there are eight RP200 diesel generators, each producing 1.75MW of power. No other aircraft carrier design in the RN has been powered by gas turbine engines and the weight saving which results from not having to use steam turbine propulsion with the associated oil-fired boilers is quite considerable.

As early as 1957 a decision was taken by the Admiralty to allow for gas turbine propulsion in warships, leading to its first introduction of gas turbines in the 'County' class guided missile destroyers, where the machinery was combined with steam turbine units. After trials in a converted frigate, the first all gas turbine ships were the commercially designed Type 21 'Amazon' class frigates, first commissioned in 1974.

The use of gas turbines not only allows for a reduction in weight and space requirements, but also of complements, effectively halving the Marine Engineering Department of the class, as compared to *Bulwark*, a light fleet carrier which carried out similar roles in the early 1970s.

The machinery chosen was the proven Rolls-Royce Olympus engine which powers the Concorde supersonic passenger transport. In its TM3B form the Olympus develops 28,000 shaft horsepower (shp) and is coupled in pairs to the twin shafts of the ship, through a David Brown reversible gearbox. The order for Olympus power was made in 1969, although at that time the

design concept of the warship was still as a through-deck cruiser rather than as an aircraft carrier. This resulted in a new approach to machinery space design; mainly because of the amount of power being generated, a conventional controllable pitch propeller would need to be specially designed, and therefore a fixed-pitch propeller was substituted. The final solution was the RN's largest propeller ever, although there were initial problems with the design causing vibration at the stern end of the ship, causing at least one propeller to be changed during acceptance trials. Even today there is an appreciable amount of shudder generated at certain revolutions.

Because of the scale of the engineering work required to produce the gas turbine engine layout in the class, Rolls-Royce built a shore test bed to represent the port-side shaft, with its two Olympus engines. The total combination was of engine, reversing gearbox, intakes, exhausts, a low-speed dynamometer and a complete set of

Right:
Developing 112,000shp the four Rolls-Royce TM3B Olympus gas turbine engines can push an 'Invincible' class aircraft carrier along at a service maximum speed of 28kt; two engines running can produce an economic cruise of 18kt giving the ships a range of 5,000nm. This is *Invincible* during naval sea trials in the Irish Sea in 1981. *DBGI*

41

ship's engine controls. Never before had a complete side of a ship's machinery been recreated ashore. This experiment proved especially successful during sea trials and later, during the passage to the South Atlantic, when *Invincible* had engine problems which resulted in a complete change of one Olympus, the experience built up by Rolls-Royce proved of great assistance in mid-ocean. Of course, complex and high technology does mean that more highly-skilled technical leading ratings are required than is necessary in steam turbine ships, but the advantages far out-weigh the disadvantages.

The Marine Olympus engine gives the 'Invincible' class a maximum speed of 28kt (52km/hr), although it is thought that 30kt (56km/hr) was achieved during trials. At the same time a long endurance is possible at the cruising speed of 18kt (33km/hr) giving a possible unrefuelled range of 5,000nm (9,260km) in normal sea conditions. The engines are interchangeable with those fitted to the Type 22 'Broadsword' class frigates and the Type 42 guided missile destroyers which usually form the ship's escort forces.

Previously the RN has used two different gas turbines for cruising and top speed, in the COGOG (COmbined Gas Or Gas) arrangement, thus almost doubling the maintenance and spares provisions. With one type of engine the whole engineering function is easier to manage,

especially with the added advantage of designed-in machinery removal trunks to the main hangar deck. This is necessary because the RN has embraced a policy of upkeep by exchange (U by E) which must be carried out away from harbour facilities at such times as operational requirements dictate.

The choice of gas turbine engine affected the internal layout of the ships, as we have seen in the last chapter, but overall the engines give the command a flexible solution to speed and endurance problems.

Rolls-Royce itself cites five major criteria for the success of the TM3B variant of the Marine Olympus:

● Its simplicity
● its reliability
● its safe and easy operation
● its ability to operate in the marine environment
● its ability to operate in wartime conditions

This latter ability proved to be most important during the 1982 South Atlantic conflict and although *Invincible* did not sustain enemy fire, six

Below:
The power behind the ships is the Olympus which has been marinised for warships, giving the Royal Navy what it considers to be the best available gas turbine in its class. This shows the engines for *Invincible* at Rolls-Royce's Bristol Engine Division before delivery to Vickers. *Rolls-Royce*

important features of the TM3B were available if the ship had been hit.

Firstly, the engines can run without electrical supplies and even retain a quick-start capability in such conditions. Furthermore, it is possible to restart the engine immediately after shut-down, which is a function of the turbine blade design (whereas in some gas turbines expansion causes the risk of blades striking their casings in such conditions). Thirdly, the Olympus is capable of taking vertical shock moments of 35g and this could be important in the event of enemy action. Also, with a self-contained oil lubrication system, the engines can run even if the ship's own supplies have been destroyed. Finally, in the event of violent manoeuvres, the engine will continue to run with negative fuel supply pressure.

In order to drive the propeller shafts, the power from the four Olympus engines has to be taken through a transmission gearbox arrangement. After initial test gear units were designed and built by David Brown Gear Industries, the Ministry of Defence placed orders with the company to supply the main propulsion gear units of triple reduction design, with fluid couplings providing 'ahead' and 'astern' manoeuvre facilities. Direct

ahead drive is through synchronous self-shifting clutches. The gearboxes are fully automatic with computer-controlled systems and a total of six have been ordered from the manufacturers at Huddersfield — two each for *Invincible*, *Illustrious* and *Ark Royal*.

The main engines may be controlled directly from the bridge and an automatic steering gear is fitted to all three ships in the class.

Falklands Conflict Statistics for *Invincible*

During the 166 days continuously at sea, *Invincible* steamed 51,660nm (95,674km) and consumed 30,196 tonnes of diesel oil for the ship's gas turbines. This can be expressed as sufficient fuel for a British Leyland Mini car, travelling at 40mph (64km/hr), to travel to the moon and back — four times!

The four gas turbines were operational for 8,011 hours, with two complete engine changes taking place at sea, creating a world record for engine changes for main propulsion gas turbines. At the same time, 18,800 Imp gallons (85,350 litres) of lubricating oil was consumed for the engines.

Invincible's daily electrical output was 72MW of power, which at British domestic tariffs for the same period in 1982, would have given a bill for £6,000, with a total equivalent bill of almost £1million for the 74 days of Operation 'Corporate'.

Below:
Before installation aboard the 'Invincible' class, the gas turbine engine concept was tested by Rolls-Royce ashore. *Rolls-Royce*

Above:
One of the COGAG triple reduction gearboxes installed in *Invincible*. Reversing and ahead manoeuvring is achieved through hydraulic couplings contained within the gearbox's after end, whilst for normal speed ahead up to full power, the drive is taken through automatic synchronising clutches. *DBGI*

Left:
An interesting view of a series of Olympus engines under construction at Bristol. Note the turbine under inspection in the middle background. Each engine unit is designed to be replaced with ease, even whilst the ships are under way. *Rolls-Royce*

Above right:
Wearing ear defenders for noise protection, an engineer officer and a petty officer (left) operate one of the four Olympus engines in *Illustrious's* engine room. This compartment complex was originally designed for steam turbine power and is consequently very spacious compared to earlier carrier designs. *Rolls-Royce*

Right:
The heart of the propulsion and associated system is the Ship's Control Centre (SCC) on one of the enclosed decks. In the modern Royal Navy, engineering watchkeeping has been made less complex by the use of high quality instrumentation to monitor the engines' performance, without the need for heavy manpower per watch to be in the actual engine rooms. This view of *Illustrious's* SCC shows the engine and shaft controls. *Rolls-Royce*

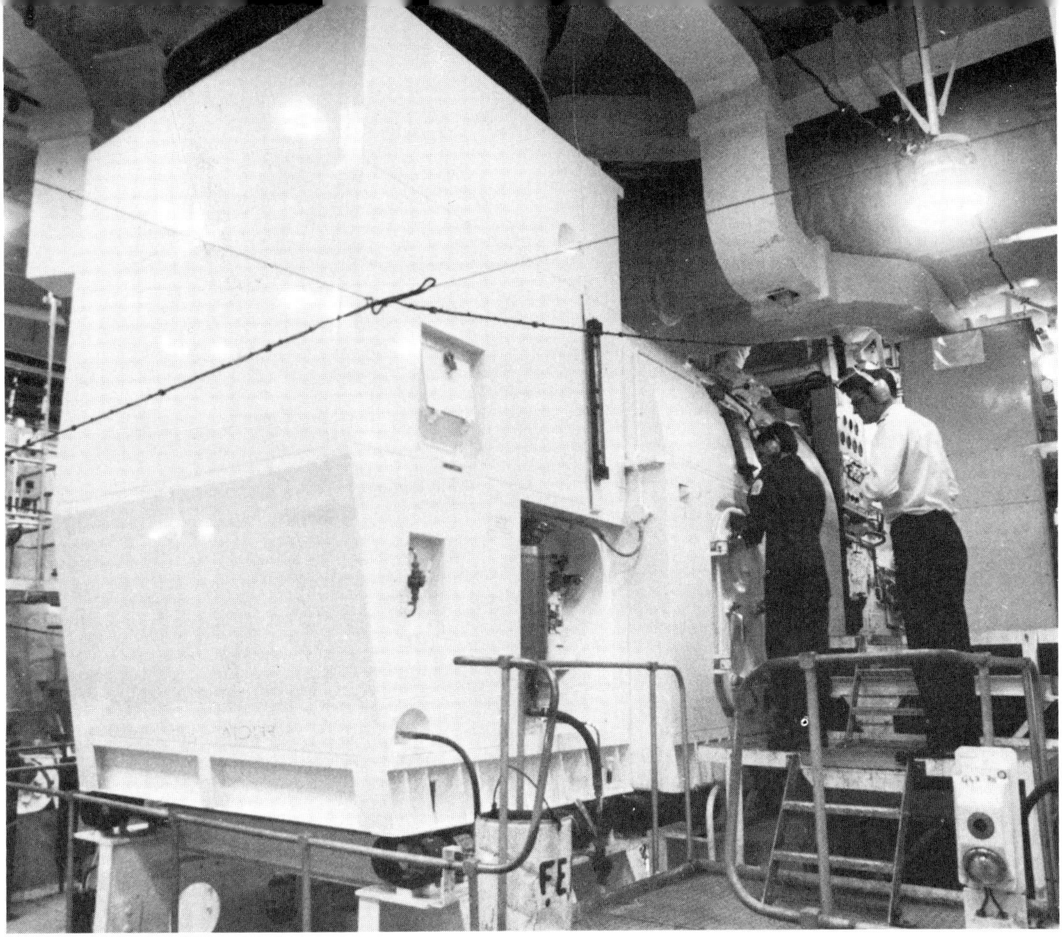

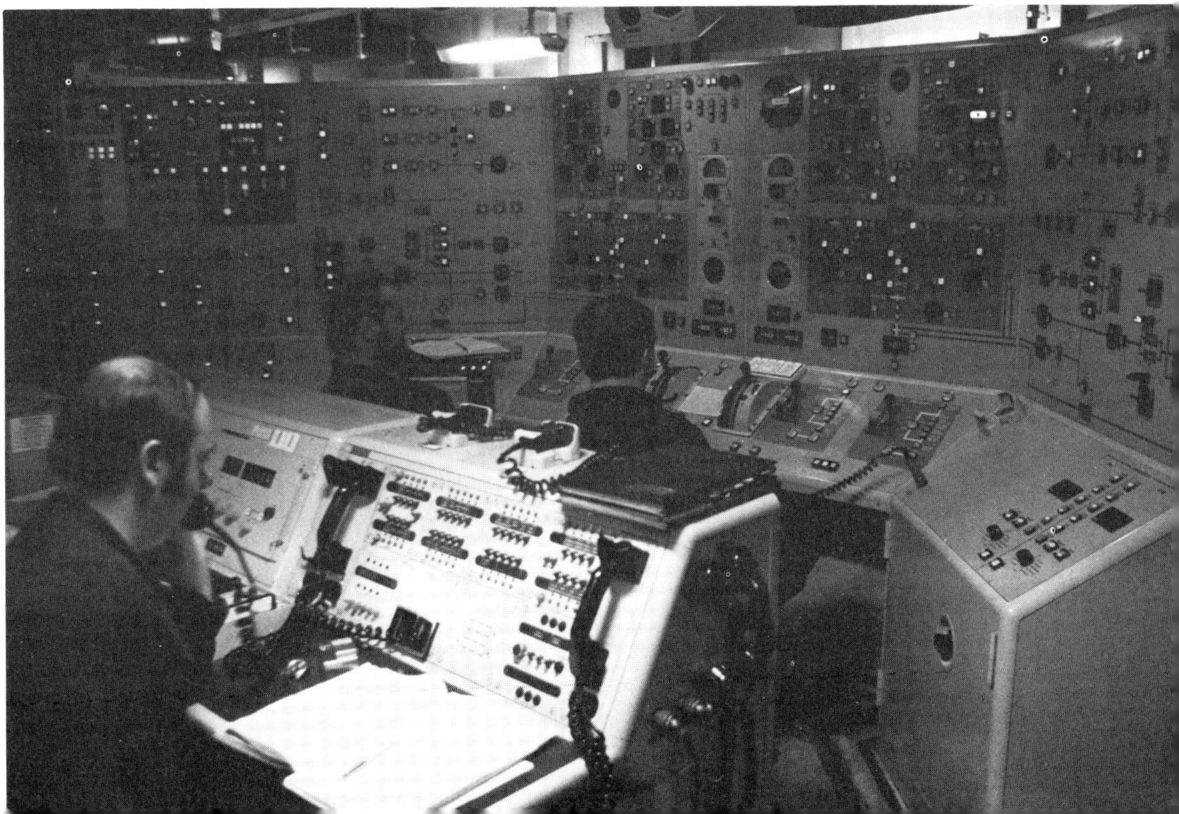

5 Sensors and Communications

As the design of the 'Invincible' class calls for a command, control and communications ship, capable of operation as *the* command ship in a task force, it is not surprising that the ships are remarkably well appointed in the C³ department.

Electrical power for the ships is provided by eight Paxman Valenta diesel generators which have specially designed covers to prevent excess noise from escaping into their compartment. The main user of the power generated is the suite of C³ compartments where a wide variety of sensors and communications systems have their control facilities and end points. The aim is to provide the Flag Officer with sufficient information, both raw and processed data, to enable him to exercise control over ships, both surface and sub-surface,

and aircraft over a wide area of sea. In addition, the ship itself has systems which themselves must be deployed and controlled.

The systems used can be broken down into radar, sonar and electronic warfare sensors, radios and satellite communications systems. At their heart is a computer complex using Ferranti FM1600 hardware, linked to the modern operations room set-up which contains a wide variety of terminals and displays, some of which are

Below:
The major surveillance and tracking systems of ***Illustrious*** **are shown in this illustration of the radar fit. All three ships of the class have the same three main radars and missile trackers.** *HMS Osprey*

A—After Type 909 (in dome); **B**—After Type 1006; **C**—Type 992; **D**—Type 1006; **E**—Type 1022; **F**—Forward Type 909 (in dome).

semi-automatic. The computer gathers the data from the various sensors, processes it and feeds it to appropriate displays in order that the situation relevant to each operator is adequately displayed.

Radar

The radars are carried high on the ships' islands because they are always hampered with the problems related directly to the curvature of the earth which limits their horizon. This limited horizon is the reason for embarking Sea King AEW helicopters in *Illustrious* when it was deployed to the South Atlantic in 1982 (see next chapter). In addition, it has been argued that this lack of far-seeing 'eyes' caused the successful attack on *Sheffield* to go undetected for so long before the Exocet missile was seen. The resuilt was a tragic day in British naval history.

Besides navigation, search, and detection, the radars in the carriers of the 'Invincible' class include two Type 909 tracker/illuminators which assist the Sea Dart missile system to acquire a target. Although the missile has a semi-active homer built into the Marconi design, the radar provides good target data until the missile is able to position itself to destroy the target. The Type 909 scanners are housed in protective radomes at either end of the island from where they can provide data to the single Sea Dart launcher, nestling behind its protective screen on the fo'c'sle. The Type 909 has a Cassegrain type antenna, with an IFF (Identification, Friend or Foe) transponder to determine whether a target could indeed be one of the ship's own aircraft, rather than a threat. The main dish has a diameter of 8ft (2.4m).

The ships of the class are also fitted with two Kelvin Hughes Type 1006 navigation radars which are commercially designed and used. This has the advantage that the radar's signature, when detected by other ESM systems, could be classified as being a commercial ship rather than a warship.

The main search radar is the Type 992R which was originally developed in the 1950s to give smaller warships a high-powered, long range radar for work to about 120nm (222km). The system has been steadily developed over the years, including the Type 992Q which provides data for the Action Information Centres of the Type 42 destroyers and the Type 21 frigates. The aerial is found high on the mainmast between the two funnels in the 'Invincible' class.

Invincible was the first ship to be fitted with the jointly-developed Anglo-Dutch Type 1022

Below:
The Type 1022 radar is a joint development with the Royal Netherlands Navy and provides Command with long range search and surveillance data. Note the radar signal-absorbent material on the foremast which prevents too much interference with the radar's performance. *Author*

long-range air surveillance radar. The British Marconi company provides the antenna, which is carried above the bridge, and Hollandse Signaalapparaten has developed the transmitter and receiver from its successful Naval Early Warning Radar LW03. Few details of Type 1022 are available, except to say that it is replacing the Type 965 'Bedstead' radar for long-range work, and has already entered service in the Type 42 Batch II destroyers.

Sonar

Besides the airborne Type 195M active sonar carried by the Sea King helicopters of the carrier's air group, and the passive/active sonobuoys, again dropped by helicopters (with the processing of data carried out aboard using the LAPADS equipment), the 'Invincibles' are also equipped with Types 184 (*Invincible* and *Illustrious*) and 2016 (*Ark Royal*).

The Type 184 is a relatively elderly active or passive sonar system, manufactured by Graseby, and mounted in a radome under the carrier's bow. In this position, it has a full 360° arc but would probably not be used in action situations for fear of giving enemy submarines the exact location of such a valuable asset.

Ark Royal is thought to be fitted with the highly advanced Type 2016 sonar which improves classification and detection through the use of the ship's AIO (Action Information Organisation). It is thought to have torpedo warning alarms in even the passive mode and trials were carried out in Type 22 frigates before the system was declared operational. It is manufactured by Plessey Marine and fitted to the ship's bow in a GRP radome.

Electronic Warfare

This sphere of modern naval warfare has only really come to the forefront in the years of the latter half of World War 2 and the postwar period. Today, most naval powers spend a good deal of time, trouble and money to prepare themselves to fight an electronic as well as a shooting battle at sea. Electronic warfare (EW) is a discipline which moves across the whole spectrum of naval systems, including radar, missile control and gunnery in particular. There are two basic areas: active EW (active jamming) and passive EW (surveillance and intelligence gathering). It should be borne in mind that this latter area is an everday occurrence at sea and the Soviets are especially adept at such work. It will not be surprising therefore if there is little to report about the 'Invincible' class EW equipment.

The most easily recognised EW devices aboard are the direction finding (DF) aerials and it was through these systems that the Allies managed to detect and therefore beat the Axis submarines in the Battle of the Atlantic. A ship listening only on

passive sets does not of course have to give herself away, but in the event of detection, active jamming can be carried out.

The problem with active jamming is that most modern missile systems have a lock-on-jam device which enables them to 'fly' straight at the jamming ship (or helicopter) if the jamming lasts long enough. In fact, jamming is carefully controlled for this reason. Modern radars and sonars are becoming 'hardened' to EW activity and the whole science of electronic warfare spends it time leapfrogging from system to countermeasure to new system.

The two most widely used terms are ESM (electronic support measures), which detect enemy radars and trackers, and ECM (electronic countermeasures), which 'attack' the enemy emissions.

Communications

This field of naval systems is regarded as highly classified, so it is only possible to give a general impression of the scope of the facilities aboard an 'Invincible' class aircraft carrier.

Main Communications Office

This facility is in the focus of the ship's external communications and uses automatic message handling machinery, considerably reducing manpower compared to ships of previous tonnage in the past. Electronic storage of signals is possible which allows them to be distributed in priority order to special terminals around the ship.

Radios

The class has an extensive fit of Ultra High (UHF), Very High (VHF), Medium and Low frequency radios to give full coverage of the ship-aircraft and ship-ship communications which are required in a modern command ship for tactical purposes. Common aerial work is used extensively. The High Frequency (HF) radio system, also used for tactical communications, includes the new and successful ICS-3 radio, made by Marconi, which has considerable export potential.

Left:
On the mainmast top, the Type 992 is used for target indication, with secondary roles for tactical air search. It is an old system, now in its 'R' modification level and it is possible that the Royal Navy will choose to replace it during the class' half-life refits. *Author*

Right:
The Type 1006 navigation radar is basically similar to that used by many commercial ships. The after set (illustrated) provides additional information for the CCA system. The two glass-reinforced plastic radomes also seen in this photograph protect the SCOT aerials from the weather. *Author*

Telebrief System

This enables aircrew to be briefed at the last minute by means of secure communications. It is a flight deck system which has its shipborne end in 'Flyco'. It is also used on all warships in the RN with flight decks and is applicable to all FAA helicopters.

Flight Deck Mag Loop

This is an electro-magnetic audio system for transmitting information to the flight deck handlers. It is also found on all warships and auxiliaries with flight decks, but only the three 'Invincible' class ships have a talk-back facility. Excessively noisy environments, like the engine room, also have Mag Loop systems.

Operations Room

Based around the Action Information Organisation concept, which enhances the facilities of the Command for, inter alia, picture compilation, threat evaluation and weapons assignment, both *Invincible* and *Illustrious* have computer-assisted operations (ops) rooms.

The AIO system, Action Data Automated Weapons System Mk 6, known as ADAWS 6, also provides facilities for Sea Dart and fighter control, and data link functions. The physical layout is based on four tactical positions — one each for the Flag Officer's staff, air warfare,

A—SCOT antennae (in domes); B—HF radio aerial; C—VHF radio aerial; D—UHF radio aerial; E—MF radio aerial; F—IFF aerial; G—HF/DF; H—UHF/DF position (when raised).

Above:
An annotated photograph of *Invincible* showing the major elements of the class's communications and electronic warfare equipment. The external fittings give little indication of the amount of gear carried below decks which help to make the 'Invincibles' some of the best fitted warships in the world. *HMS Invincible*

Right:
Capt The Hon Nicholas Hill-Norton explains the functions of one of the four positions in the operations room to Rear Adm Vallings, Flag Officer Gibraltar. *HMS Invincible*

undersea warfare and surface navigation. Surrounding these positions are the different consoles for directing the various facets of the ship's weapons and sensors, including air direction, countermeasures, missile firing, Phalanx gun control, electronic warfare (EW), sonar and CCA (carrier controlled approach) radar. In both *Invincible* and *Illustrious* CCA is carried out through the medium of the Type 1006 radar, using the antenna on the after end of the superstructure. This set will be replaced or supplemented when the ships are taken in hand for their major refits, and facilities made available for the MEL MADGE approach aid. *Ark Royal's* fittings are thought to be already in place.

Ark Royal's ops room will have basically the same layout as the first two ships in the class, but has the larger and more advanced ADAWS 10 and is considered to be the culmination in both sophistication and size for the shipborne centralised computer. There are small differences in the actual layout of the ops room in the earlier ships, and on *Illustrious* the Flag Staff planning space contains additional display facilities relating to the ship's larger communications fit.

Basically then, the ops room is the focus of the

sensor and weapons management, forming an integrated command facility. The class can, of course, be fought without the Flag Staff aboard and, indeed, during the Falklands conflict *Invincible* was used as the anti-air warfare and fighter direction ship for the CVBG, having a more comprehensive AIO fit than the flagship, *Hermes*.

Computer Room

The power behind the ops room and its ADAWS system comes from the provision of two Ferranti FM1600 computers which give real-time facilities and capacity. Unlike a straight data processor, the real time computer used has a priority system for the evaluation of data.

The computers provide two functions for ADAWS integration; firstly, they allow the presentation of information for decision making, and secondly, they allow data links between ships. In fact, the three 'Invincible' class ships are the only ADAWS ships at present capable of sending data between systems, using a telex-type system. The picture compilation facilities allow for an enlargement of the surface, air and subsurface information between units, via data link, to give the flagship a very wide picture. This is more than would be normally possible with one ship's radar alone.

Ship's Inertial Navigation System
The system fitted to the 'Invincible' class consists of three gyros which measure the ship's acceleration across the earth's surface, integrated for distance and speed. The system, similar to that carried in nuclear-powered hunter-killer submarines (SSNs), requires only details of the start point to give a position, accurate to a few hundred yards, anywhere in the world. It also provides real time latitude and longitude information and assists with weapons stabilisation. SINS is backed by a Mk 19 Sperry gyro compass, which also can be used to provide weapons stabilisation information.

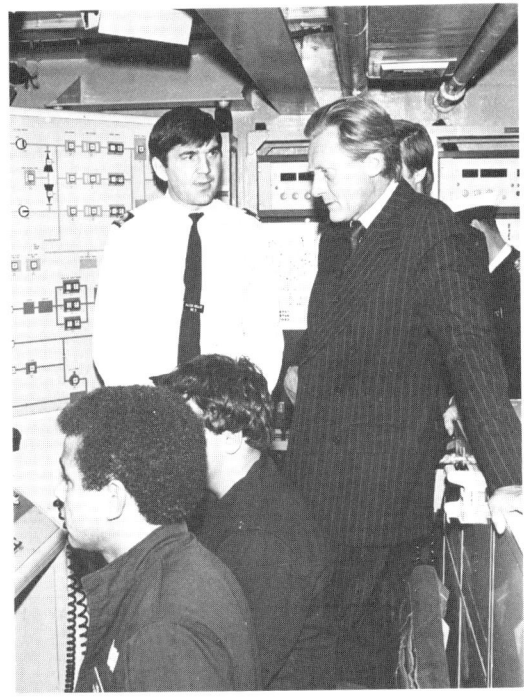

Right:
The Ship's Control Centre, positioned immediately adjacent to HQ1, the Damage Control Centre, is perhaps the heart of the ships of the 'Invincible' class. This picture shows an interested visitor, Michael Heseltine, Secretary of State for Defence, being briefed on the layout of the SCC by Sub Lt Allen Hollis, *Invincible's* **Flight Deck and Hydraulics Engineer.** *HMS Invincible*

6 Weaponry

Although it was not intended to be so, the aviation complement of the 'Invincible' class, particularly the Sea Harrier jets, has become the main weaponry of the carriers. The original command cruiser idea envisaged some helicopter assets being carried for anti-submarine duties, both in support of a task group and to defend the parent ship, but air defence and anti-ship threats would be dealt with by the Sea Dart missile system, or, more probably, by the escort frigates and destroyers forming the carrier's screen. By the middle 1970s the class had become an aircraft carrier design to embark the revolutionary Sea Harrier 'jump-jet' which could more readily provide area air defence and surface strike. In the early 1980s the helicopter elements had been additionally strengthened by the introduction of the Sea King HAS5 and the Sea King AEW.

British Aerospace Sea Harrier FRS1

In 1958, the Hawker company (now essentially part of the nationalised British Aerospace Corporation, but then a private company) began the development of vertical take-off and landing (VTOL) aircraft. The design was to be powered by a single engine with vectored nozzles to give forward speed, and rotated downwards for lift. The engine chosen was the Bristol Siddeley Pegasus which underwent flying trials in 1963 — a growth version of this same Pegasus powers the Sea Harrier today.

Interest was shown in the project by the Royal Air Force, Admiralty and, through NATO, by the United States and Federal Germany. This resulted in trials with the derivative Kestrel aircraft and eventually to a design contract for a two-place supersonic VTOL fighter for naval use, known as the P1154. This aircraft would have Mach 2.5 performance and be powered by either two Rolls-Royce turbofans, or a single Bristol-Siddeley BS100 turbofan. Although twin-engined aircraft are normally favoured for carrier operations, no real decision was made about the engine fit for some time. By 1964, the office of the Director of Naval Air Warfare had been studying the design in some detail and had come to the conclusion that the American F-4 Phantom would actually suit British requirements better, bearing in mind the CVA-01 proposal was still very much alive. The P1154 did not in any case come up to its expectations.

In the period 1964-66 the new Labour Government set about a series of unsympathetic defence and technology-related cuts which, as has been mentioned already, led to the cancellation of the CVA-01 and the P1154 (amongst other projects). The RAF was able to reconstruct its Harrier programme however and began pursuing a single-seat, single-engined ground attack and strike fighter for tactical operations on NATO's Central Front in Europe. A contract was signed in 1967 and the first production aircraft flew just after Christmas 1967, with RAF evaluation the following year. By 1969 the first RAF squadron had the Harrier GR1 in service, to be followed by the current GR3 in 1976. In 1970, however, the RAF took its Harriers to sea in *Eagle* to acquaint itself with a flight deck, clear the aircraft for maritime flying and show the FAA a VSTOL (Vertical and Short Take-Off and Landing) aircraft. The FAA was impressed and invited the squadron back the following year, when the Admiralty began a Staff study into the design.

The Naval Staff Requirement proposed in general terms the following three basic roles, which are described below, together with the associated mission profiles.

Primary Role

Air interception and fleet fighter, which involves the tracking and interception of long-range maritime patrol aircraft, including those providing mid-course guidance for anti-ship missile systems, and for anti-shipping strike aircraft posing a threat to the task group. The Sea Harrier's radius of action for this role is 400nm (741km) at 20,000ft (6,096m), during which time it would be fully armed with AIM-9L and 30mm Aden cannon. In the combat air patrol (CAP) role the Sea Harrier is able to remain on station for one and a half hours, with a 100nm (185km) transit to and from the patrol area.

Secondary Role

Reconnaissance and probe in which 20,000sq nm

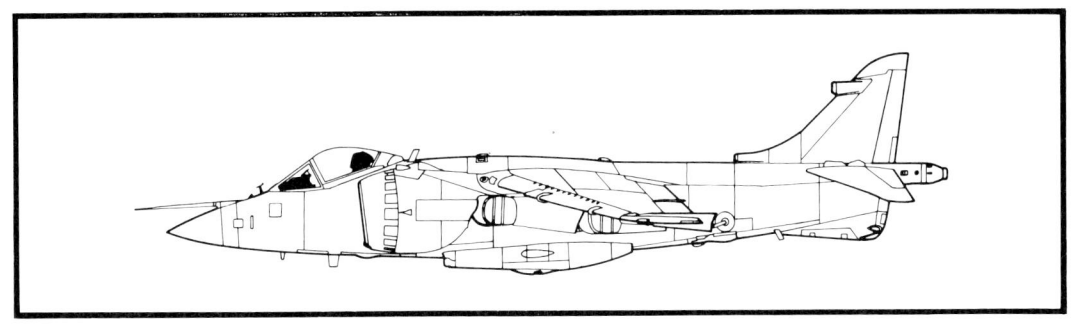

(59,623sq km) could be covered in one hour flying time at low level. The probable target of such a role would be surface and/or air threats, as well as the gathering of visual and tactical intelligence of enemy positions and responses.

Above:

The Sea Harrier FRS1 is the main air defence system open to the 'Invincibles' to defend themselves and their task groups. These two aircraft from the temporarily re-formed 809 Squadron, recover aboard *Illustrious* during its deployment in the South Atlantic. The nearer Sea Harrier is flown by Lt Cdr Tim Gedge, the CO of 809 Squadron. *HMS Illustrious*

Tertiary Role

Strike and ground attack, which was demonstrated during the Falklands conflict and, dependent on armament and fuel tank fits, would involve 250nm (463km) transits to a target area and return.

The RAF Harrier design needed to be changed to give important resistance to salt-water corrosion and improved landing gear modifications. The cockpit area was essentially redesigned to give the pilot — a single pilot, which was unusual by modern naval fighter standards — a better all-round view for combat manoeuvring. To assist the

single pilot flying the three roles envisaged, the avionics needed to be improved to give Head-Up Display (HUD) information and air-to-air combat data, an attitude reference platform which was self-aligning, and the Ferranti Blue Fox radar with its TV-type display.

Complementary to the AIM-9L Sidewinder and 30mm cannon already mentioned, the Sea Harrier would be fitted with Cluster bombs for the ground attack role and Iron bombs for bombing sorties. Later it was decided, as a result of the experience in the Falklands, to up the standard two

Sidewinder fit to four missiles carried in pairs on underwing stations. In addition, the British Aerospace Dynamics Group P3T Sea Eagle missile was added to the inventory for anti-ship operations. Besides the Blue Fox system, passive sensors included a radar warning receiver, carried in the tail. Normal radio and transponders were also specified. If all else fails, the Sea Harrier is now equipped with a Martin-Baker Type 10 zero-zero ejection seat (which is quicker-acting than the RAF type).

Basically, to navalise the Harrier seven magnesium components had to be eliminated from the new airframe and two more major components from the Pegasus engine. For deck use the Sea Harrier would require tie-down lugs, an emergency braking system, increased positive tailplane travel, a new electrical system and a simple autopilot. All these elements only raised the aircraft's total weight by less than 100lb (45kg) — this has been described as an achievement unequalled in the transposition of a modern aircraft to a naval environment.

A production contract was signed rather late — May 1975 — for 34 Sea Harrier FRS1 aircraft to embark in the 'Invincible' class aircraft carriers, with the first unit commissioning on 19 September 1979. After operational trials in *Hermes*, the first front-line unit (800 Squadron) was deployed on the same ship, with 801

Squadron, the second Sea Harrier squadron, going to *Invincible* after formation in February 1981.

The Sea Harrier went into action in 1982, being embarked in both *Hermes* and *Invincible* during their deployment to the South Atlantic around the Falkland Islands. Here the aircraft distinguished themselves by taking on superior numbers of enemy aircraft and not losing a single aircraft to air combat. The period was one of boredom for the crews, punctuated by bouts of action which they had never experienced before; *Invincible's* operations room teams did not have an opportunity to look outside for eight weeks during this time.

Sea Harrier pilots have praised the way in which the carrier worked during the whole period of Operation 'Corporate', and some put this down to the training exercises in the previous spring off northern Norway where the weather was generally worse than around the Falklands. This meant that the ops team, aircrew, controllers and

Below:
For a short while 800 Squadron operated from *Invincible* **(as signified by the letter 'N' on the tip of the aircraft's tail). Note the normal Aden cannon pack has been replaced by a cluster bomb carrier in this picture.** *HMS Heron*

flight deck personnel had had the opportunity of working in conditions which did not make the South Atlantic a shock to the system. Some pilots had to spend 10 to 14 hours a day in the ready room, whilst in general each of the Sea Harriers embarked from 801 Squadron flew three times their normal flying rate and each aircraft is reckoned to have completed about 80 hours' flying during the 74 days of the conflict. Yet the quickest scramble was made in just 2min 35sec. Even when the hostilities ashore ceased, the Sea Harriers were still on alert, making a grand total of four and half months in this state of readiness.

When it ventured south to relieve *Invincible*, the newly accepted *Illustrious* embarked on 809 Squadron, which had flown down to Ascension Island and moved by ship to the Total Exclusion Zone (TEZ) around the disputed islands. 809 Squadron disbanded on its return to the UK and duties in the two 'Invincible' class carriers will be shared by 800 and 801 Squadrons, although it is possible that a third unit will be formed in 1985.

Westland Sea King HAS5
The Sea King design is based on the Sikorsky S-61 or SH-3 helicopter which was developed in the United States, but has been produced under licence by Westland Helicopters in the UK since the late 1960s.

The Westland Sea King HAS1 entered service in 1969 as the RN's primary anti-submarine helicopter, being embarked in the 'Tiger' class cruisers, *Ark Royal* (IV) and the commando carriers, in the latter on an occasional basis. As the design developed in service, the basic engineering was improved, especially in terms of the power plants and transmission. Today, the Sea King has passed through the HAS2 stage and is operational in 'Invincible' class aircraft carriers, *Hermes* and Royal Fleet Auxiliaries, as well as being land-based for both training and for guarding the approaches to the Clyde submarine facilities. By 1984 all previous Mk 2 helicopters will be retro-fitted with improved avionics to the Mk 5; in addition, there is a quantity of new-build helicopters, some to replace battle and attrition losses in the Falklands conflict.

As well as operating in the ASW role, the Sea King can be employed on Search and Rescue missions, trooping flights, logistics and stores transfers (Vertrep), and helicopter delivery service (HDS) work. All Sea Kings have a basic crew of two pilots, an observer and an aircrewman. The latter two crew members are accommodated aft in the darkened rear cabin with the Sea Searcher radar and underwater situation displays; the ob-

Below:
Sea Harrier weapon fit.

A—100Imp gal combat tank; **B**—Twin AIM-9L Sidewinder launchers; **C**—190 Imp gal ferry tank; **D**—Practice bomb carrier; **E**—LEPUS flare dispenser; **F**—1,000lb Iron bomb; **G**—28lb Practice bomb; **H**—30mm Aden cannon; **I**—4lb Practice bomb.

server, usually the aircraft's commander, operates the main display, whilst the aircrewman (a rating) operates the sonar set and the processor equipment. For those Sea Kings equipped with passive sonobuoys, the aircrewman is responsible for their operation, usually released through the cabin door. During SAR operations, a specially trained diver might also be carried and the aircrewman becomes the winch operator.

The first HAS5 model was handed over to the FAA on 2 October 1980 at RNAS Culdrose. It first reached operational service with 820 Squadron, which embarked in *Invincible*. Amongst the pilots joining the unit was HRH Prince Andrew who, as a Sub Lieutenant, flew operational combat sorties with the ship during the Falklands conflict.

In 1982 814 Squadron, also equipped with the Sea King HAS5, embarked two in *Illustrious* on 24 June to test the aviation facilities. Later the rest of the squadron embarked, together with two Sea King AEW (see below) from D Flight of 824 Squadron.

The following statistics, provided by 820 Squadron, indicate how important the Sea King

HAS5 was to the success of the South Atlantic Task Force:

Aircraft embarked: 9
Hours flown: 4,700
Sorties flown: 1,650 (including passive ASW, HDS, exercises with friendly submarines and surface search/ESM)
Average hours per pilot: 321
Average hours per observer: 336
Loads lifted: 1,172
Stores lifted (tons): 1,200
Torpedoes dropped: 6
Depth charges dropped: 10

The flying statistics could be put another way: the squadron could have flown around the world twice, and the average flying time represented the same numbers of hours as if five aircrew had been airborne for the whole period of *Invincible's* 166 days at sea. This was in itself a world record for continuous aircraft carrier operations.

Westland Sea King AEW

Although, as we have seen, the Westland-built

56

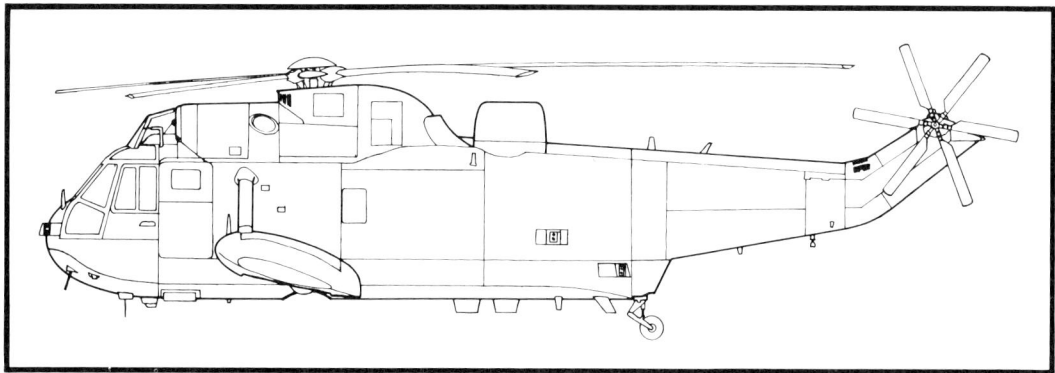

Above left:
The use of the ski-jump ramp has greatly extended the abilities of the nonetheless remarkable Sea Harrier. Here, during deck qualifying trials — known as Sea Acceptance Trials Air (SATAIR) and, a later series, called First of Class and Controller Aircraft Trials (CA) between July and November 1980 — a Sea Harrier launches from *Invincible*. *RN*

Top:
The main aim of the 'Invincible' concept is to put to sea a large number of ASW helicopters. The Westland Sea King HAS5 is the latest development in medium ASW helicopters and currently equips all the front-line squadrons embarked in British carriers, Royal Fleet Auxiliaries and converted container ships. Sea King 012 is from 820 Squadron in *Invincible* and is pictured dropping a Mk 46 ASW torpedo. *HMS Invincible*

version of the S-61 anti-submarine helicopter is no stranger to the flight decks of the 'Invincible' class, the hybrid AEW (Airborne Early Warning) model is a relatively new development.

In the spring of 1982, the UK, and more especially the RN, found itself with a possible war situation some 8,000nm (14,816km) from home — almost half a world away. It is now a well known fact that the Fleet found itself vulnerable to low-flying air attack, mainly from stand-off missiles launched from strike aircraft. The sinking of the Type 42 destroyer *Sheffield*, built mainly with the intention of giving area air defence to 'Invincible' class carrier task groups, proved that the reliance on surface-carried radar for air warning was unworkable at best and positively dangerous at worst.

Just 12 days after the loss of *Sheffield*, on 16 May Westland Helicopters, together with Thorn EMI, the manufacturer of the Searchwater radar already fitted into the Hawker Siddeley Nimrod MR1 maritime reconnaissance aircraft, delivered a completed design study to the RN. The subject of this study was the installation of a modified Searchwater system, together with all associated systems, into a modified Sea King HAS2 airframe for AEW duties with the Fleet. Actually, Westland had already carried out private venture studies into the feasibility of mounting a radar into a helicopter during 1968, a year before the Sea King entered RN service. The work needed in 1982 was nevertheless intensive. By early June the RN — particularly the FAA — had reviewed the situation and work was ordered to

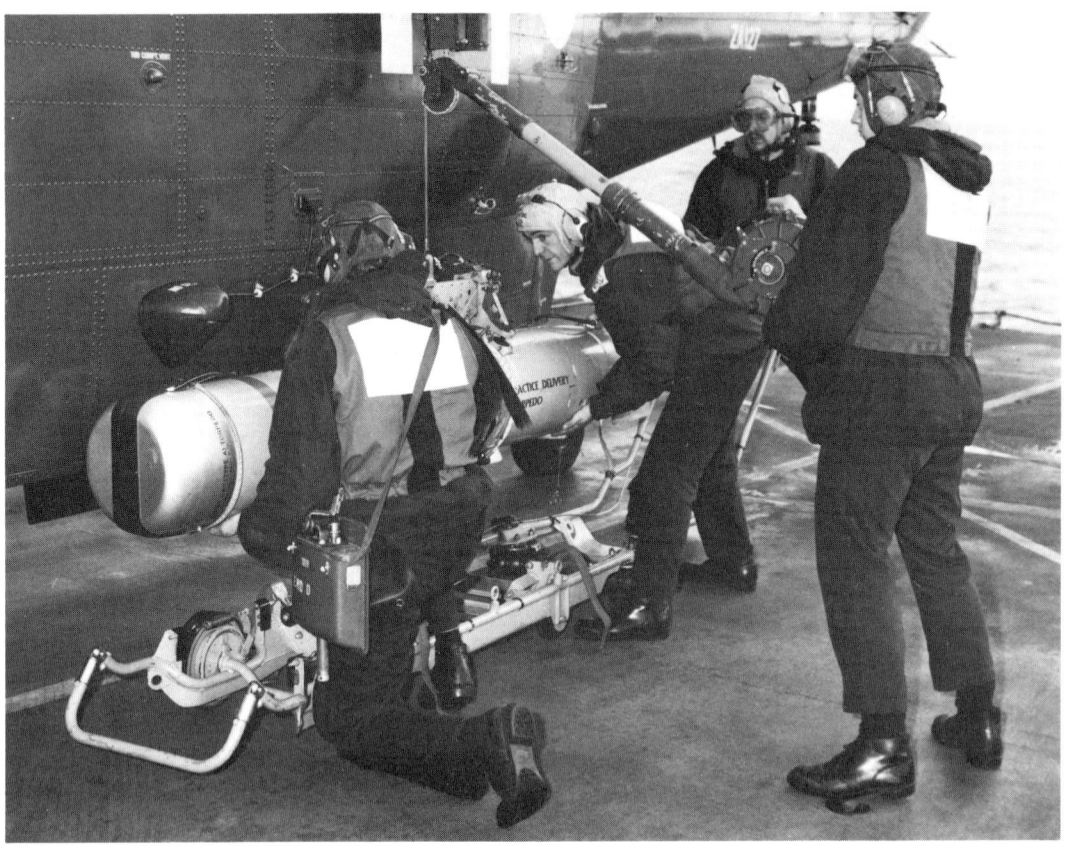

commence, using two HAS2 airframes made available by RNAS Culdrose.

As we shall see in the next chapter, the speed with which the men at Swan Hunter's yard had completed the fitting out of the second of the class, named *Illustrious*, meant that the two AEW trials Sea Kings had a full size deck on which to operate. On 2 August the helicopters and their Searchwater radar had been thoroughly tested and they were embarked to *Illustrious* in the Channel. In less than three months, British industry had effectively produced a new aircraft with a new role. There were apparently sufficient numbers of RN observers who had flown the Navy's last AEW aircraft, the fixed-wing Gannet AEW3 from *Ark Royal* (IV, decommissioned in 1978), and so a flight could be formed to take the helicopters south. It was planned that *Illustrious* and its air group, including the two Sea King AEW helicopters, would go south and relieve *Invincible* in the immediate post-hostilities period.

After several months away in the South Atlantic, the now designated D Flight of 824 Squadron returned to Culdrose having proved that the radar set could detect targets at ranges in excess of 100nm (185km), a considerable improvement on radars closer to the surface. In

Above:
A Mk 46 practice delivery torpedo is loaded aboard a Sea King helicopter during 'Springtrain '82' exercises prior to the Falklands and South Atlantic deployment of both *Invincible* and 820 Squadron. *HMS Invincible*

Right:
The conversion of two Sea King HAS2 airframes to take the Thorn-EMI Searchwater radar was one of the major triumphs of the Falklands campaign — on the home front. Actually, the helicopters did not arrive in the South Atlantic in time to see active service, but the concept appears set to stay in the Royal Navy. The radar antenna is housed in the inflatable radome seen in its deployed position in this photograph. *RNAS Culdrose*

fact, it is thought that the Sea Kings carry the Searchwater to approximately 10,000ft (3,048m) during operational sorties.

The radar itself is fitted into the after cabin of the Sea King, with the antannae kept in an inflatable radome on the starboard side of the fuselage, which this is hydraulically swung into position shortly after take-off but not activated until the operational area is reached. On returning to the ship, the radome is swung up 90deg to the stowed position again. There are no blind areas with Searchwater, it being a full 360deg radar which also allows for fighter direction duties to be carried out by the helicopter, either with the 'Invincible' class' own Sea Harrier fighters or with friendly carrier or shore-based aircraft.

The Searchwater and associated equipment, including the Racal Avionics Electronics Surveillance Measures (ESM) antennae, gives the Sea King AEW a gross weight of 21,400lb (9,714kg), as against 20,500lb (9,298kg) for the ASW versions. Although the RN intends to have further AEW helicopters of this type in service, the fact that D Flight missed 'Orient Express' deployment to the Far East suggests that all is not well with the system in some way. When fully operational it will however give the RN an important edge in detecting low flying threats.

EH Industries EH101

By the end of this decade, the Westland Sea King will be a rapidly ageing design and work has already begun on a replacement helicopter capable of fulfilling the roles already carried out by the Sea King, and even enhancing them. Although it will probably first be assigned to the new Type 23 escort frigates, the Anglo-Italian (Westland-Agusta) EH101 helicopter will be deployed to the 'Invincible' class aircraft carriers.

The design is a three-engined, high technology helicopter with anti-submarine warfare as a primary mission, but apparently plans are now well in hand to develop an AEW version for sea service in not only the three 'Invincibles' but also

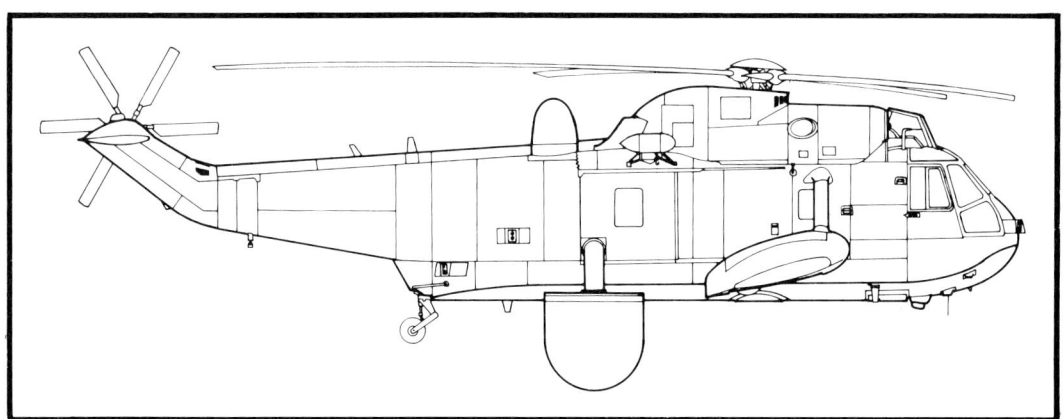

Royal Fleet Auxiliaries. It is planned that the helicopter will be able to operate in 98% of the weather found in the North Atlantic and be capable of fully autonomous operations day or night in wind conditions approaching 60kt (111km/hr). The planned endurance is up to five hours in the role of searching for enemy submarine contacts at about 50nm (93km) from the carrier, and having found a target, the helicopter systems will classify it, process the data and begin an attack with the Marconi Stingray lightweight ASW torpedo.

Unlike the present Sea King, the EH101's radar will be housed in an under fuselage radome; the radar favoured is the Ferranti Blue Kestrel development of the Sea Harrier's Blue Fox and Lynx' Seaspray, giving 360deg coverage and track-while-scan capability. In addition to ESM gear, the cockpit arrangement will include the latest technology in colour displays giving aircraft performance, attitude and other data. Being three-engined the helicopter will be safer than any twin-engined type and this will be important for SAR (Search and Rescue) operations from the carriers. The first flight is expected in 1986 and service entry in 1989.

Other Aircraft

From time to time, operational requirements dictate that other types of helicopters must be carried in 'Invincible' class ships. During exercises with Royal Marine Commandos, the helicopter support is usually supplied by Sea King HC4s (846 Squadron) and Wessex HU5s (845 Squadron). In addition, during its work-up period, *Illustrious* embarked a Wessex HU5 from 772 Squadron (Portland) to provide HDS and transportation as well as SAR facilities.

The ships are capable of operating Harrier GR3 ground-attack fighters of the Royal Air Force, and during the Falklands conflict the aircraft were embarked for a short period. Other helicopter types are also capable of operating from the ships' decks.

Below:
In July 1982 *Illustrious* embarked these two Sea King AEW helicopters for its deployment in the South Atlantic. They are pictured coming back to the ship after a trials flight with their radomes stowed. More helicopters of the type are on order to give the RN the ability to place three three-aircraft flights in the 'Invincible' class carriers by 1985. *HMS Illustrious*

Sea Dart Guided Missile System

The Sea Dart system was designed by British Aerospace Dynamics Group to provide medium-range area air defence for a naval task force and was first fitted to the Type 82 destroyer *Bristol* in 1973. Since then GWS30, as the RN denotes the complete missile system, has been fitted to the Type 42 destroyers and the three 'Invincible' class aircraft carriers.

Modern Soviet aircraft pose a threat which is centred around a massed high, medium and low level saturation raid against a naval task force with several types of aircraft, depending on the targets' position relative to shore bases. The Sea Dart is thought to be adequate for this type of attack, having been proved against silver-foil covered balloons at sea level and high level during a simultaneous salvo firing of two Sea Dart mis-siles by a Type 42 destroyer. The missile uses semi-active homing and is powered by a Rolls-Royce ramjet motor to give it Mach 2.5 performance. During the Falklands conflict *Invincible* launched six Sea Dart missiles in anger in less than two minutes, on 25 May, and this is believed to be greatest number of missiles fired by an RN ship in such a short time. This was the day that the ship was believed to have been the target for a raid of Argentine Navy Super Etendard fighters, armed with Exocet AM39 missiles; in the event, the container ship *Atlantic Conveyor* was sunk and neither *Invincible* nor *Hermes*, the flagship, were hit.

Sea Dart also has an out-to-the-horizon anti-ship role which has yet to be tried in anger. The missile is 'flown' by hydraulically-powered actuators at the rear of the missile. The four pro-truding dielectric antennae determine the orientation of a target and direct the missile by pro-portional navigation to an intercept; the antennae are called 'polyrods' in the RN.

Illustrious and *Ark Royal* also are fitted with a single Sea Dart launcher amidships, the former proving its system during July 1982 in the Portland Sea Training Area just prior to sailing to join the South Atlantic Task Force.

Below:

An early impression of the EH101 project which aims to provide a new, three-engined helicopter for naval use by 1992. The helicopter will almost certainly equip the naval air squadrons embarked in the 'Invincible' class carriers and is a joint production venture by Westland (UK) and Agusta (Italy). *Westland Helicopters*

Above:
Although not part of the regular air groups, Sea King HC4 helicopters have operated aboard both *Invincible* and *Illustrious* during commando exercises. They have a support role and this example from 846 Squadron is seen with a light trailer as an underslung load. *Author*

20mm Oerlikon GAM – BO1

There has been much and varied discussion in the media and in other published works concerning the lessons learned by the Royal Navy during the South Atlantic conflict with the Argentine in April-June 1982. Although this book is not a vehicle to continue such discussion, it is pertinent to record that following the initial air attacks on British warships by Argentine strike aircraft, the completion of *Illustrious* at Wallsend-on-Tyne was slightly delayed whilst gun armament was fitted. At the builder's yard, as we shall see below, a revolutionary close-in weapon system was fitted, but at Portsmouth the ship received one of the oldest weapons still in operational service with the Royal Navy — the 20mm Oerlikon cannon.

The Oerlikon GAM-BO1 type gun is a robust and high performance mounting, using the Oerlikon 20mm belt-fed automatic cannon, capable of firing 1,000 rounds per minute, with a good supply of ready-use ammunition of varying types available. This type of weapon was first readily identified aboard a British warship when the Prime Minister, Mrs Margaret Thatcher, visited the 'County' class guided missile destroyer *Antrim* at San Carlos shortly after hostilities ended. Two guns are fitted on the upper deck of the carriers' superstructure island, just forward of the second funnel (or uptake). Both *Invincible* and *Illustrious* had been fitted by early 1983, and *Ark Royal* was given the light and simple mountings during fitting out. Although of Swiss design, the 20mm Oerlikon KAA is built in the UK by the British Manufacture & Research Group of Grantham, Lincolnshire, under licence from its parent company, Oerlikon-Bührle.

20mm Vulcan-Phalanx

The threat of sea skimming anti-ship missiles to RN surface vessels became a reality in May 1982 with the sinking of *Sheffield*, and later the destruction of *Atlantic Conveyor*. Before the Falklands conflict ended, *Glamorgan* had also been attacked and hit but not destroyed by an Exocet missile. These weapons were not the latest available and so the RN was doubly spurred into fitting a counter to them.

Illustrious was fitted with such a counter from the builder's yard, and on arrival at Portsmouth on 21 June many were surprised to see two units of a close-in weapons system fitted to the open

When the decision was made to embark Sea Harrier jets and to incorporate a ski-jump ramp, the single Sea Dart launcher was removed to starboard. As can be seen from this view from the bridge of a High Seas Firing of a live Sea Dart, the amendment has not altered the firepower. *HMS Illustrious*

fo'c'sle and the starboard quarter of the flight deck. The 'Daleks' had arrived in the Royal Navy. The carrier had been fitted with the American CIWS Mk 15, of Vulcan Phalanx 20mm Gatling-type rapid fire guns. They are fully automatic and controlled by integral radar which allows them to engage both supersonic aircraft approaching the ship, and sea-skimming missiles. They are, in effect, a last ditch defence. Following successful firing trials on 20 June, the system was declared operational after further trials in the English Channel on 26 June 1982. In fact, during the first firing of a Phalanx gun, a low level Rushton target was destroyed with the gun's first burst of fire.

When *Illustrious's* sister-ship, *Invincible*, returned to Portsmouth in September 1982, it was taken in hand for Phalanx fitting, although in

Right:
Another close-range system is the Oerlikon GAM-BO1 20mm mounting which can be used against aircraft, missiles or surface targets. It is a powerful weapon, manufactured in Britain and sold throughout the world.
HMS Invincible

future the RN may not use the system on the Type 42 destroyers as originally thought. In addition, the fitting of the CIWS Mk 15 to *Ark Royal* is also in doubt at this time.

Countermeasures

The conflict in the South Atlantic showed that a considerable amount of faith could be placed in countermeasures using Chaff, to distract and generally 'put off' incoming missiles. Chaff is the modern equivalent of 'window' dropped during World War 2 by aircraft trying to disguise their position with German radar controllers, thus preventing accurate interception. Today, warships and auxiliaries carry Chaff dispensers of several different types. The 'Invincible' class uses two:

3in Rocket Launcher System
Formerly called the Corvus launcher, *Invincible* and *Illustrious* have an arrangement of 16 tubes immediately aft of the main mast. These can train automatically and be used to fire several types of countermeasures material. They have been supplemented by:

Super Rapid Blooming Onboard Chaff (SRBOC)
This is an American/Canadian system designed to act as a countermeasure to incoming missiles by 'seducing' them away from the ship.

Left:
The Phalanx CIWS was purchased 'off the shelf' from the United States Navy which has fitted the weapon extensively in its larger warships. This view of *Illustrious's* forward Phalanx appears to indicate the gun under local control, but it can just as easily operate automatically. Note the UHF/DF aerial which folds to starboard when not in use. *HMS Heron*

Below left:
A close-up of the Phalanx mounting showing, from the top, the acquisition and tracking radar radome, the 20mm Vulcan Gatling cannon and the continuous feed drum-type magazine. *HMS Invincible*

Right:
Missiles can be deterred, seduced and otherwise foiled in their attempts to home on to a ship, by the use of chaff. This is the 3in rocket launcher mounting (port side), which used to be called the Corvus launcher. When not in use, as pictured in *Invincible*, the mounting has its muzzles covered for protection against the elements. *Author*

Below:
Following the Falklands conflict, *Invincible* was fitted with American SRBOC launchers which also deploy chaff to defend the ship against incoming enemy missiles. The mounting can be fired locally or from special locations in the ship, including the operations room. *Author*

A—Forward Phalanx; B—SRBOC launchers; C—Two 20mm Oerlikon GAM-B01 mountings; D—After Phalanx.

Above:
An indication of the additions made to the defensive systems of *Invincible* after the Falklands conflict: Phalanx CIWS, SRBOC launchers and Oerlikon GAM-B01 guns. *Illustrious* was completed with these features. *Ark Royal* in its turn has a different arrangement of defensive systems. *HMS Invincible*

Below:
Illustrious as completed with Phalanx forward and aft, plus the new single Oerlikon mountings aft of the mainmast on the superstructure's 02 deck. This is the ship exercising in the English Channel during June 1984. *Author*

7 In Service

HMS *Invincible*

The name *Invincible* is not new to the Royal Navy. In fact, there have been six ships bearing the title in the last 200 years.

During 1982, when *Invincible* was one of two aircraft carriers in the South Atlantic operating in support of the British Forces which were liberating the Falkland Islands, the fact that the fifth *Invincible* had been one of the leading ships in the Battle of Falklands of 1914 did not escape the crew. In December 1914 the German cruisers *Scharnhorst* and *Gneisenau* were engaged off the Falklands by *Invincible* and *Inflexible*, when the former was the flagship of Vice Adm Sir Doveton Sturdee's squadron in the South Atlantic. On returning to home waters *Invincible* took part in the Battle of Jutland as flagship of the Third Battle Cruiser Squadron, but was sadly sunk with loss of more than 1,000 of the ship's complement.

Some 230 years to the day of the present *Invincible*'s launching, Adm Anson captured the French 74-gunner of the same name off Cape Finisterre to start the name in the annals of the Royal Navy. Between 1747 and 1758, when it was wrecked off Portsmouth, this ship served the Crown off the American coast.

The second *Invincible* was also a 74-gunner, built at Deptford. It took part in many glorious actions, including Capt St Vincent (1780), St Kitts (1782), and the First of June (1794), but was wrecked off the East Anglian coast whilst hurrying to join Adm Lord Nelson in the Baltic. The three battles mentioned are now Battle Honours, along

Below:
During Sea Acceptance Trials, *Invincible* was visited by a Sea King HAS2 helicopter from 706 Squadron, the training unit at RNAS Culdrose, and by a Sea Harrier FRS1 from the Intensive Flying Trials Unit — 800A Squadron — at RNAS Yeovilton. *RNAS Culdrose*

with Alexandria (1882) in which the fourth *Invincible* took part. Heligoland (1914), plus the Falkland Islands and Jutland, already mentioned, were battles in which the fifth *Invincible* earnt Battle Honours.

When *Invincible* sailed into Portsmouth Harbour on a very dark and grey day in March 1980, it was the largest warship built for the Royal Navy in 25 years. The ship was accepted by the Royal Navy on 19 March 1980 after successfully completing 'first of class' sea trials off the West Coast during which time the major systems were put through a number of rigorous and important tests. Particularly important were the tests for the four Rolls-Royce Olympus gas turbine engines and their associated equipment. At the time, the power shafts, the propellers and the electrical power generation systems were the largest then built for the Royal Navy. *Invincible* also had the largest air conditioning system afloat in Britain and this was an important aid to the closed down 'citadel' situation used during NBC (Nuclear Biological Chemical) operations.

During the initial trials there were some doubts about the ship's propellers and vibrations caused at high speed. The propeller problems were cured after Part IV sea trials in the Channel and the successful work-up with a full RN crew. The vibration problem is still occasionally apparent and was particularly noticeable during the Falklands campaign when several of the aircrew complained about the vibration making it impossible for them to get enough sleep. After about two weeks of operations, the aircrew learned to sleep anywhere! Needless to say, the following two ships of the class have been redesigned to overcome the problem.

When commissioned on 11 July 1980, *Invincible* had the distinction of being the world's first ship with a STO (short take-off) ramp — or ski-jump — and of being the first with the unique, three-side-loading scissor lifts. The two lifts, in the conventional forward and after positions in the hangar, allow for the easier movement of aircraft, and the three-side loading allows, for

Below:
On 5 April 1982 *Invincible* followed *Hermes* out of Portsmouth, heading south for Ascension and the South Atlantic. The South Atlantic Task Force was sailing to war some 8,000 miles away. On *Invincible's* flight deck there is a line-up of Sea Harriers from 801 and 899 Squadrons together with Sea King HAS5 helicopters from 820 and 706 Squadrons — all still in peacetime markings. *RN*

example, the repositioning of aircraft struck below should one go unserviceable. The lifts are also more battle-damage proof.

The use of the gas turbines for all the power requirements gave *Invincible* another record — the largest gas turbine-powered ship afloat.

After the commissioning and a period of summer leave, *Invincible* continued to work up, particularly with the Sea Harriers of 801 Squadron and the Sea King HAS2s (later HAS5s) of 820 Squadron. The former unit had only re-formed in February 1981, and joined the ship in May. There then followed continued work-up and operational readiness exercises, including the participation of the ship and its air group in large NATO exercises of the year, 'Ocean Venture' and 'Ocean Safari'. During this period, 801 Squadron flew over 1,000 hours from *Invincible*, which was declared operationally ready for service on 16 June 1981.

It is interesting to note that during 'Ocean Safari' the Sea Harriers of the ship were able to maintain an adequate CAP (combat air patrol) for task force operations for more than 80 hours and during this period denied the force's air space to all intruders. The ship and 801 Squadron proved that it was possible to launch, via the ramp, four Sea Harriers in 50 seconds, during which time one of the accompanying US Navy nuclear-powered aircraft carriers could only launch one F-14A Tomcat. The ability for rapid launch and the ability to keep aircraft in the air were decisive factors in the Falklands campaign less than a year later. In addition, during the 1981 exercise season it was found that *Invincible* could continue to operate aircraft in sea conditions which would have prevented the conventional, catapult-launched jets of the US and other navies from taking to the air. Considering that the ship spent, by its own figures, 31% of the sea time in the South Atlantic in sea state 5 or greater conditions, this ability to launch aircraft, not necessarily into wind, is most useful and certainly proved the 'Invincible' concept; if indeed the concept of Sea Harrier and light carrier needed to be proved.

In late 1981 *Invincible* conducted a full trial in its tertiary role of amphibious platform for Royal

Below:
Invincible at Ascension Island on passage to the Falklands. The opportunity was taken to transfer stores and test equipment before the CVBG actually deployed 'south' to liberate the Falklands. *HMS Excellent*

Marines Commandos of the United Kingdom/ Netherlands Amphibious Force, part of the NATO reinforcement for the Northern Flank in the event of tension or war. Of particular importance during this trial was the ability of *Invincible* to use its speed to transit to the Amphibious Objective Area by means of a 'quick dash'. This could be from Arbroath, the headquarters of 45 Commando, RM, or to southern Norway and collect forces there. In the latter connection it should be noted that the UK/NL force spends at least three months of the year in Norway for training and exercises.

In March 1982, during Exercise 'Alloy Express' in the northern Norwegian Sea, *Invincible* operated with a full air group of five Sea Harriers and nine Sea Kings (for ASW) as well as embarking a reduced Commando and its support helicopters, the latter being part of 846 Squadron flying Westland Sea King HC4 Commando helicopters and normally based at Yeovilton. During the exercise's amphibious phase, the commandos were landed by air having spent their time aboard using the hangar for accommodation. Whilst landing the men, the ship's integral air group continued to fly CAPs and ASW screens around the task group. This

Top:
The Commanding Officer during the operations in the South Atlantic was Jeremy Black — known as 'JJ' to the ship's company of *Invincible*. Following his success in fighting *Invincible* during the conflict, he was promoted Rear Admiral and in September 1983 joined the ship again as the flag officer for the 'Orient Express' deployment. *HMS Invincible*

Above:
During the early stages of Operation 'Corporate', Sea Harriers, like these from 801 Squadron, were armed

with 1,000lb Iron bombs to attack Argentine positions. The pilot climbing into the centre Sea Harrier's cockpit is Lt Cdr Nigel 'Sharkey' Ward, CO of 801 Squadron at the time. *HMS Excellent*

Right:
Arming the Sea Harrier with an AIM-9L Sidewinder air-to-air missile. These weapons proved far superior to anything which the Argentine Air Force or Naval Air Arm could muster and have been credited with the destruction of 27 enemy aircraft. *HMS Excellent*

experience was so useful to the ship and the Royal Marines when the situation in the South Atlantic called for the sending of the Task Force. *Invincible*, then fully worked up, had been at sea for varying periods during the previous 12 months and was perhaps in the best condition possible for war service, except that the ship itself was in need of an assisted maintenance period (AMP) at the very least. In fact, during the voyage to the Total Exclusion Zone (TEZ), the ship's company achieved a remarkable feat in changing one of the four Olympus engines.

With great feeling, crowds at Southsea witnessed the departure of the bulk of the Task Force, including *Hermes*, the flagship, and other warships on 5 April 1982. *Invincible* had a full warload of eight Sea Harriers embarked, the latter having flown up from Yeovilton earlier. During the voyage south the augmented Sea King squadron (still 820) and the Sea Harriers continued training and work-up until they were ready for action by the time *Invincible* reached Ascension Island, after a passage lasting nearly three weeks. The time at Ascension was very short, allowing only for the transfer of the commandos to other ships and some replenishment of the ship. The Sea

Harriers were unable to complete their training ashore and so had to fly once more from the deck on the way to TEZ.

Despite its own defensive armament of a twin Sea Dart missile launcher, Rear Adm John 'Sandy' (now Sir John) Woodward, the Task Force commander, wisely chose to keep frigates in close attendance to *Invincible* to act as 'goalkeepers'. These warships were usually the Type 22 'Broadsword' class frigates which were armed with the Sea Wolf GWS25 missile in sextuple launchers fore and aft. *Invincible's* Sea Darts needed the back-up of the latter because they were purely area defence weapons. The ship did use its missiles 'in anger' on 25 May, when six were launched in less than two minutes. This was also the day that the *Atlantic Conveyor* was sunk, as well as being the national day of the Argentine, hence the heavy air raids against the fleet.

Interestingly enough, no Sea Darts were reportedly fired on 30 May, the day on which the Argentine Air Force and Navy claimed to have launched a devastating attack on the carrier with A-4 Skyhawks and a single Exocet-carrying Super Etendard. In fact, according to Argentine sources, the priority targets for the fighter-bombers were

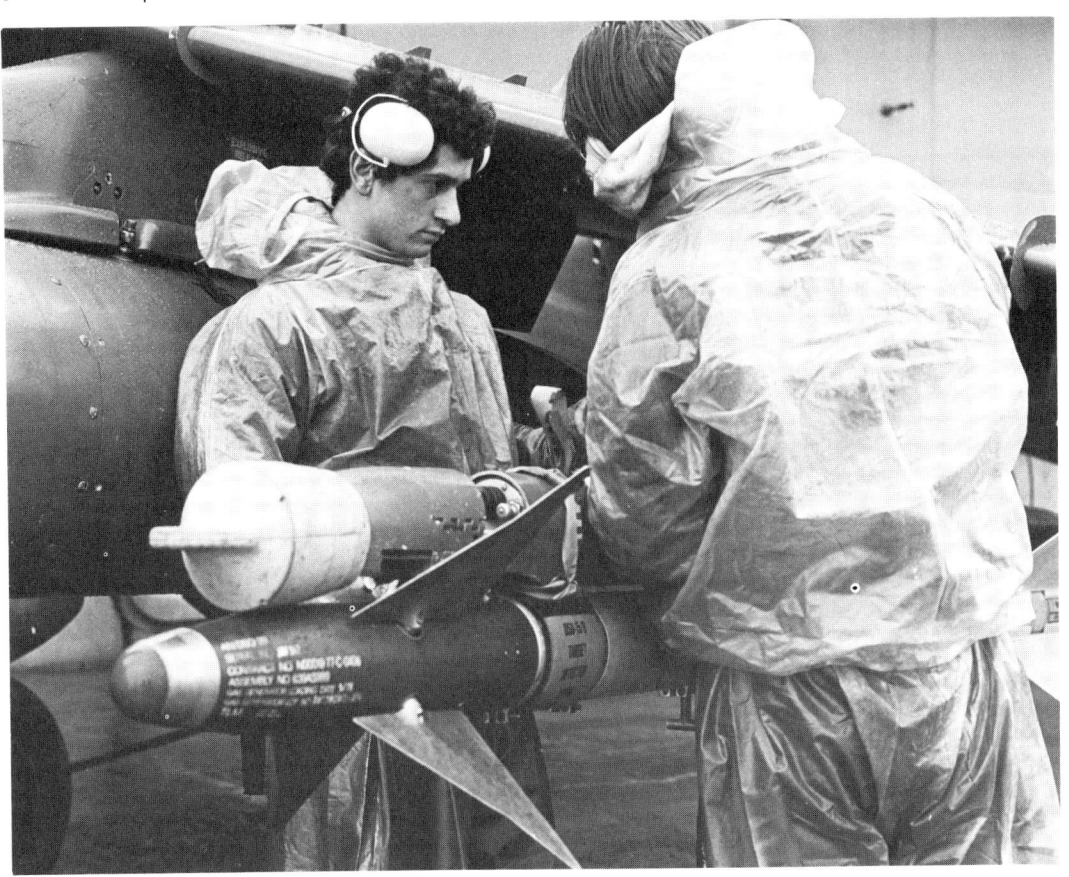

Above:
Although not designed as a bomber, the Sea Harrier, like this one from 801 Squadron, can carry 1,000lb Iron bombs (illustrated), BL755 Cluster bombs and other underwing stores for strike missions. Note also the 30mm Aden cannon and 100 Imp gal combat tanks.
HMS Excellent

the two carriers. The attack was apparently mounted using four A-4C Skyhawks, each carrying two 500lb (227kg) bombs, which used two Super Etendards as navigation leaders. In addition, one of the 'Supers' was carrying the last remaining AM39 Exocet missile in the Argentine inventory. After refuelling from a Boeing 707 tanker aircraft (formerly flown by the Argentine state airlines), the attacking aircraft found their target which was bombed by the A-4s and hit by the Exocet. Although the Argentine is quite adamant that the attack was successful, it has not been possible to find any sources to confirm that *Invincible* was closely attacked, let alone hit, on that day, or any other during the conflict. There does appear to have been an Exocet fired in the direction of the *Invincible* group but it exploded nowhere near any warship or accompanying merchant vessel. The Argentine Navy (and several others who should know better) claim that *Invincible* was hit by a missile in one of its quarters and that two of the A-4Cs achieved hits as well (the remaining two being destroyed by the escorts' Sea Darts). They claim that *Invincible* remained in the South Atlantic until September in order that repairs and repainting could be carried out for the triumphant return to Portsmouth on 17 September 1982. There is certainly no indication of any damage to the ship, nor has there ever been.

During the Falklands conflict *Invincible* remained at sea for 166 days, this being a world record for continuous aircraft carrier operations, and during this time it 'steamed' a total of 51,660nm (95,674km), or the equivalent of twice around the globe. As a result of the ship's travels, it used 30,196 tonnes of diesel fuel, and the aircraft, plus those of visiting flights of squadrons, consumed a total of 7,620 tonnes of aviation gasoline. The ship's impressive fresh-water distillation plant, which can also provide sufficient water for the troops in an overload situation, manufactured 8.6million gallons (39million litres) of fresh water for drinking, washing, cooking and other purposes.

In addition to the Olympus engine which was changed at sea during the passage south, another engine was replaced after the conflict, but before the ship returned to Portsmouth. The former occasion was the first time in world maritime history that a ship has changed an engine at sea.

The air group was also busy during Operation 'Corporate', for a period of 45 days at war and

263 hours at Action Stations. The Sea Harriers of 801 Squadron (augmented where necessary) flew 1,580 hours which amounted to 1,430 sorties, these being divided into 950 combat air patrols, 300 strike and bombing missions and 180 reconnaissance sorties over both land and sea. 801 Squadron was responsible for the action which resulted in the capture of the Argentine spy trawler *Narwhal* on 9 May, when the Squadron CO flew top cover for the assaulting helicopters from *Invincible* (820 Squadron) carrying SBS troops, the trawler having been sighted by a Sea Harrier from 800 Squadron temporarily aboard *Invincible*.

As a matter of interesting statistics, 801 Squadron destroyed seven enemy aircraft in combat and three more most probably did not return to their bases on the mainland. Although not all their duties were actual combat sorties and the pilots also spent hours on Deck Alert, one pilot remained in his aircraft for an incredible nine hours. On average, each pilot flew combat for 73 hours, the longest sortie being $1\frac{3}{4}$ hours, the shortest only 10 minutes.

Sadly, one Sea Harrier was lost to enemy action during a strike mission to Goose Green, although seven enemy Pucaras were destroyed on the ground. In addition, two Sea Harriers were lost without trace and a third exploded without cause after take-off. In heavy seas and during a ship's manoeuvre, a fourth Sea Harrier slipped over the ship's side but the pilot managed to eject and was recovered by one of *Invincible's* Sea Kings within minutes.

The Sea Kings themselves were busy setting records and flew a remarkable 4,700 hours with nine helicopters. The sorties flown, which totalled 1,650, were for the Task Group's anti-submarine screen in most part, but also for helicopter delivery services, surface search and exercises with friendly forces on the way down and during the uneasy period immediately after the Argentine surrender on 14 June 1982. Operation 'Corporate' did not finish until 12 July when the Argentine Government agreed to discontinue , but still no peace treaty has been agreed by it.

820 Squadron has achieved a special fame during the Operation 'Corporate' period and since, because one of the Sea King pilots was HRH Sub Lt Prince Andrew. Despite his royal station,

Below:
Besides providing an excellent anti-submarine screen around the CVBG, the Sea King helicopters (820 Squadron illustrated), also acted as casualty evacuation aircraft. Here the Squadron's and ship's aircraft handlers and flight deck personnel carry an injured man to *Invincible's* sick bay. *HMS Excellent*

Above:
Local aircraft defence for *Invincible* during the Falklands was provided by various small arms, including this bridge-top 7.62mm L4A4 machine gun position manned by members of the ship's company. *HMS Excellent*

he flew his turn of operational mission and appears to have been airborne for Exocet seduction sorties.

The squadron statistics also give the amount of stores moved to and from *Invincible* and other ships, as 1,200 tonnes in 1,172 lifts. In action, the Sea Kings released 10 Mk 11 depth bombs and six Mk 46 homing torpedoes against possible submarine targets.

Following the cessation of hostilities and the return of *Invincible* to Portsmouth dockyard for a well-earned refit period, for both ship and crew, interesting disclosures about the efforts made by the Argentine forces to sink the ship were published. Several newspapers around the world carried the story of the attack on the aircraft carrier, whilst it was steaming in the Total Exclusion Zone, by a conventional submarine. This boat, one of two purchased from Federal German yards in the 1970s, was particularly quiet and is thought to have been in the vicinity of the CVBG during the sinking of *Sheffield*. During this time,

several independent naval sources reported seeing torpedo tracks, or hearing them on the sonar. Apparently the submarine *San Luis*, a Type 209 armed with German SST4 quiet torpedoes, moved into a firing position but, according to the Argentine, all four torpedoes fired failed to sink the carrier because although they struck the hull, they failed to explode. Research undertaken by the author, with members of the ship's company at the time of the *Sheffield* attack, or when the *Atlantic Conveyor* was hit (another possible time for a submarine attack, with all the confusion caused by mounting a rescue operation), shows that nobody recalls any underwater noises or alarms which were attributed to submarine attack. The Navy will admit that *Invincible*, and many other ships in the Task Group, took evasive action when possible submarine or torpedo contacts were detected, but at no time did *Invincible* come under attack.

Certainly, the carrier returned to Portsmouth alive and well, despite this apparent action and the previously mentioned Exocet and/or bombing attacks. Many have questioned why the ships stayed down south for so long after the end of hostilities, but as we shall see below, *Invincible* needed to provide air defence for the reoccupa-

Above:
According to Argentine sources, *Invincible* was nearly sunk on several occasions when their coastal conventional submarine *Salta* came within torpedo range. The submarine's commander was only thwarted in his attempt by faulty torpedoes, says the Argentine Navy. *HDW Kiel*

tion of the Islands until *Illustrious* could join the Task Force and take over. It was even said that the ship had been taken to Simonstown, South Africa for repairs! Nevertheless, many believe that Argentine submarines were the greatest risk to the carriers in Falklands waters.

In 1983 *Invincible* commenced the testing of one of its tertiary roles, that of the 'quick dash' troop transport and command centre: it could be that Royal Marine Commandos (including elements of the UK/NL Amphibious Force) would need to be moved with headquarters and equipment to northern Norway in support of the United Kingdom's obligations under the Atlantic Alliance's policy of defending member states against aggression.

The size of a force embarked depends on the number of Sea King ASW helicopters and/or Sea Harrier jets embarked at the same time, but up to one Commando-size unit (900 men approximately), plus a detachment of support helicopters,

can be carried for short periods. The ship had conducted a full trial in 1981 to test the feasibility of the amphibious 'quick dash' and then took part in Exercise 'Alloy Express' in northern Norway in March 1982. Although not equipped with landing craft like *Hermes*, *Invincible* has the ability to operate helicopters for trooping and supplying equipment and stores to the shore. During these exercises however the ship not only operated ASW and air defence duties with a full air group of Sea Kings and Sea Harriers embarked, together with approximately 400 air group personnel, but also succeeded in landing a full Commando group by support helicotper. Using *Hermes*, the concept of integrated ASW, air defence and commando operations was tested during Exercise 'Cold Winter', thus enabling *Invincible* to proceed with full scale amphibious-only operations for Exercise 'High Tide' (with Royal Netherlands Marines) in May 1983, followed by Exercise 'Rough Diamond' in the Solent.

'Rough Diamond' is the biannual field training and logistics exercise for 3 Commando Brigade, Royal Marines, which was widened to allow Commodore Amphibious Warfare, Cdre Peter Dingemans DSO, to exercise naval warfare on passage to the landing area. All good practice for

Above:
Invincible at sea off the Falklands on 12 June 1982 —
the day that *Glamorgan* was hit by an Exocet, but a day
of less flying than normal for the carrier. Ready to
launch are two Sea Harriers of 801 Squadron, whilst
the lighter coloured aircraft are from 809 Squadron,
formed especially for the Operation 'Corporate'
deployment. *HMS Invincible*

Left:
At peace again, **Invincible** at Gibraltar during Exercise
'Springtrain '83' in April, with the ship's air group's
aircraft ranged on deck for leaving harbour.
Mike Lennon

Above right:
In 1983 *Invincible* exchanged its Sea Harrier/Sea King
air group for Royal Marines' support helicopters from
845 and 846 Naval Air Squadrons. Illustrated is a
Sea King HC4 (846 Squadron) landing to collect another
stick from 45 Commando which was also embarked in
the carrier. *HMS Invincible*

Above, far right:
Capt The Honourable Nicholas Hill-Norton assumed
command of *Invincible* in March 1983. His father,
Admiral of the Fleet Lord Peter Hill-Norton was Chief of
the Defence Staff when *Invincible* was ordered in
1973. *HMS Invincible*

possible hostilities in Norway or any other location around the world. For the exercise, the first of its kind with 'Invincible' class carrier, helicopters from 845 Squadron (Wessex HU5s) and 846 Squadron (Sea King HC4s) were embarked.

After returning to its home port of Portsmouth, *Invincible* made ready for summer leave, which was to be followed by a hectic period as the ship prepared to take part in a major deployment to the Far East. Before setting out on the Far Eastern trip, *Invincible* was one of the stars of the Portsmouth Navy Days over the August Bank Holiday and attracted many thousands of visitors. Within a week of the Navy Days, the ship departed Portsmouth for the beginning of a period of $7\frac{1}{2}$ months away, but beginning with a period of exercise work-up on passage down the Channel. The full story of the 'Orient Express' deployment is told in chapter 8.

HMS *Illustrious*

Illustrious is also a famous 'fighting ship' name in British maritime history, with a pedigree stretching from 1789, when a third rate frigate of 74 guns was launched at Buckler's Hard, Hampshire for RN service against the French. In 1795 the ship was responsible for the capture of two French warships, but then it suffered severe damage in action, although winning the first of the 11 Battle Honours attributed to the name: Genoa 1795. The first *Illustrious* was burned in the Mediterranean.

The name was not dormant for long and was next awarded to another third rate frigate, launched in 1803 and with a career lasting 65 years, including the winning of Battle Honours for Basque Roads (1809) and Java (1811).

At the close of the Victorian period, battleship design altered radically with the introduction of the 'Dreadnoughts'. This caused *Illustrious* (III) to become obsolete almost overnight, even though it had been launched only as recently as 1896. It served World War 1 in the humble role of ammunition store ship and was scrapped in 1920.

The next *Illustrious*, built at Barrow, can never be said to have had a humble career, for the fleet aircraft carrier which bore the name was arguably one of the 10 most famous warships of World War 2 with no less than eight Battle Honours. These began with Taranto (1940) when it launched 20 Swordfish torpedo bombers to cripple the Italian Fleet in harbour. After continued service in the Mediterranean (1940-41) and on the Malta Convoys (1941), *Illustrious* was

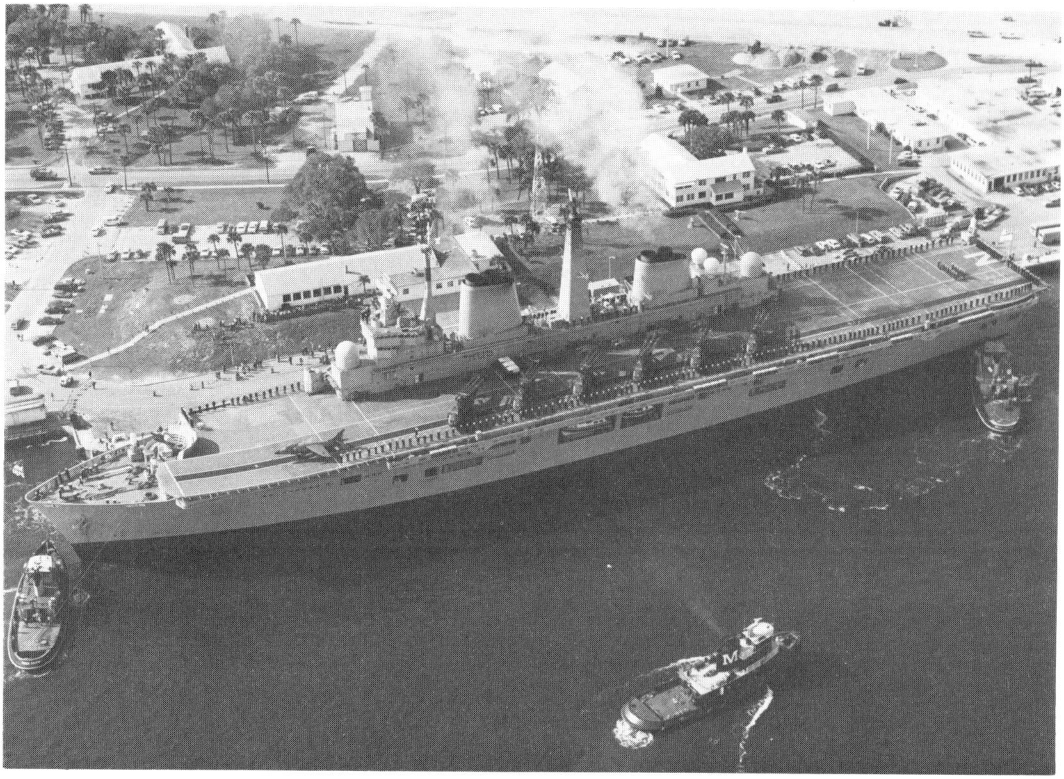

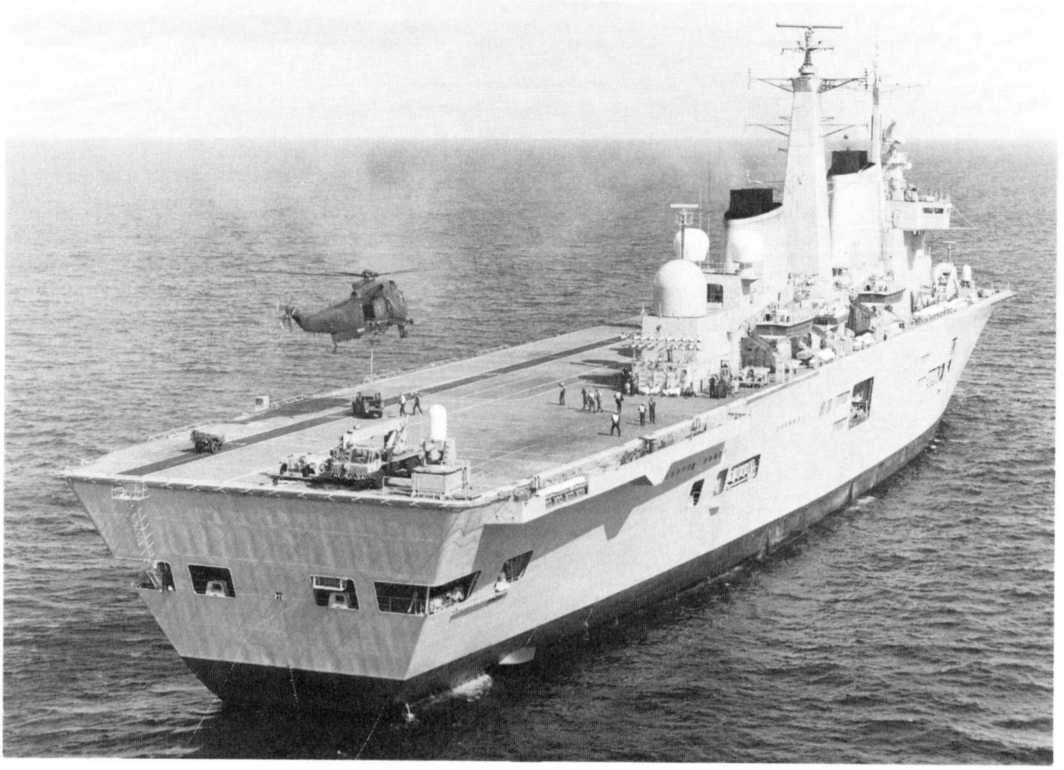

badly damaged by German air raids and had to be withdrawn to refit in the United States. There it nearly became the command of Capt Lord Louis Mountbatten, but before work was completed, he was promoted to take charge of Combined Operations. The carrier's service then continued through the Indian Ocean landings and the first probes at Europe through Italy, giving the Battle Honours Diego Suarez (1942) and Salerno (1943) respectively. There then followed service in the Indian Ocean with actions against Sabang (1944) and Palambang (1945) on the way to join the British Pacific Fleet, where the ship finished its war service during the drive towards Japan — Okinawa (1945) was *Illustrious's* last Honour.

After the war, the ship was employed in training and trials duties, but was laid up when the new generation of angled-deck aircraft carriers (*Ark Royal, Eagle* and the 'light fleets') came into service. *Illustrious* (IV) was sent to the breakers in 1957, and so passed a fine ship.

The events in the Falkland Islands and the South Atlantic generally caused the name of the fifth *Illustrious* to be brought before the public gaze rather sooner than the Admiralty had expected. Although not due to enter service until 1983, the 'Invincible' class' second member was accepted into service a year earlier and despatched to the South Atlantic. The original

Above left:
Trips to the United States are always enjoyed by British warships, to Fort Lauderdale not the least. This is *Invincible* **departing after a rest and recreation visit following the 'Caribtrain '83' exercise programme.** *RN*

Left:
Designed to be able to carry a Royal Marine Commando for 'quick dash' landing or reinforcement operations, the 'Invincible' class has experimented and exercised in the role. This is *Invincible* **during Exercise 'Rough Diamond' during 1983.** *HMS Invincible*

Above:
The completion of *Illustrious* **by the Tyneside yard of Swan Hunters was an amazing feat of willpower, trust and enterprise. The ship's company of the carrier was delighted and is seen here paying its respects to the shipyard. In early May 1983, the ship made a 'Thank You' visit to Newcastle and over 30,000 visitors toured the ship.** *Swan Hunters*

plan called for acceptance in September 1982 and trials with weapons and aircraft until the middle of 1983, when the warship would be ready for commissioning. In the event, the acceptance ceremony took place on 18 June with the ship still alongside at Swan Hunter's fitting-out berth at Wallsend-on-Tyne. Three months ahead of schedule, *Illustrious* was ready for sea and the crowd of well-wishers were proud of the North-East's achievement.

Above:
On passage to Portsmouth, in the North Sea *Illustrious* carried out some rapid acceptance trials. Basic SATAIRs were carried out in the Channel during late June, when Lt Cdr (later Cdr) David Ramsey, RAN, completed the first Sea Harrier launch from the deck in the Sea Harrier illustrated. *HMS Heron*

Illustrious then left Tyneside and proceeded to sea for trials and a rapid shake-down in the North Sea. On 20 June, a Sunday, the ship was commissioned at sea in what is probably a unique event in the history of the RN. But the ceremonies did not delay the carrier's arrival at Portsmouth the next day when an immediate bottom inspection was arranged and Harbour Acceptance Trials began. At the same time, the warship was stored with all the necessary items for war service in the South Atlantic, a process which took three days to complete.

The aircraft began to arrive on 24 June, when a plane guard Wessex HU5 from 772 Squadron joined with two Sea Kings from 814 Squadron, the units which were to provide the anti-submarine helicopters for the deployment. A Sea Harrier arrived to test the ski-jump and Lt Cdr David Ramsey, Royal Australian Navy, was the first pilot to use the deck, making, coincidentally, his 100th launch in the Sea Harrier.

As we have seen in a previous chapter, the weapons firing was successfully completed by 26 June, making it only nine days since the ship had been accepted from the builders. Perhaps another record!

After Operational Sea Training under the watchful eye of Flag Officer Sea Training's staff at Portland, *Illustrious* was ready for the South Atlantic. This was especially remarkable because of the number of newly joined personnel and because the ship itself was so new. By 2 August, admittedly after hostilities had ceased on the Falklands but still in a period of intense naval activity, the carrier was ready to deploy.

It was considered necessary to relieve *Invincible* on station in order that the latter's crew, which had spent a record time at sea (see above), should get some well-deserved rest. Capt J. C. K. 'Jock' Slater took *Illustrious* out of Portsmouth, past HMY *Britannia*, with HRH Princess Alexandria aboard, on passage for Gibraltar for stores and mail. The only unusual event was the loss overboard of a young rating and his football. This was only a temporary loss because the well exercised 'Man Overboard' drill went very smoothly and he was collected, with ball held high as a signal in the rough seas, in just four minutes. *Endurance*, the ice patrol ship which had become a household name in April, being the only RN warship in the South Atlantic when the Argentines invaded, was passed with due ceremony and four days later *Illustrious* crossed the line and proceeded to Ascension Island.

The busy roulement-type operations which always seem to herald the arrival of ships at this tiny Atlantic island were carried out and the Sea Kings and Wessex HU5s ashore (a flight from 845 Squadron) embarked 84 underslung loads for the ship and the forces ashore in the Falklands. The two sister-ships met at sea on 27 August when *Invincible* was relieved in the carrier battle group (CVBG) by *Illustrious* — almost 10 weeks to the hour after the latter had departed from Tyneside. On 28 August the ships deployed into close formation for a steampast which was featured in almost every newspaper and TV programme in the English-speaking world. Both carriers steamed close together and their crews lined the sides — many have described this event as one which they will never forget. After the steampast, Rear Adm Derek Reffel (now Vice Adm and Flag Officer Naval Air Command), Flag Officer Third Flotilla and commander of the naval forces in the South Atlantic, transferred to *Illustrious*, with his staff.

The ship's primary duty was to provide air cover with the Sea Harriers of 809 Squadron, under Lt Cdr Tim Gedge DSC, AFC, who had previously taken the Squadron south by air and later in the ill-fated *Atlantic Conveyor*. The area for protection was now called the Falkland Islands Protection Zone (FIPZ) and here *Illustrious* provided air co-ordination using the facilities of its most up-to-date air direction room and the services of the two Sea King AEW helicopters. During this period the ship remained at Defence Stations. The time was used to test the helicopter AEW system in a period of about eight weeks, until 21 October when the ship departed for the UK. High operational times were achieved and the system was used to detect low-flying aircraft in a series of tests. Basically, the Sea King enables the detection time of any threat to be increased, especially if the helicopters are deployed towards

Below:
Invincible and **Illustrious** meet for the first time — a 'steam by' off the Falkland Islands as the former was relieved on station by the latter. **Illustrious** with the light-coloured Sea Harriers of 809 Squadron is nearest the camera. *HMS Heron*

Top:

On the 'war station', *Illustrious* **is replenished at sea from the fleet oiler** *Olmeda,* **which is simultaneously replenishing the oil bunkers of the Type 21 frigate** *Amazon.* **Aircraft on** *Illustrious's* **flight deck include** *Olmeda's* **Wessex HU5 (forward at Fly One), two Sea Harriers and a Sea King HAS5 parked by the island, four further Sea Harriers aft on the start and run up positions, but most interesting of all, a Sea King AEW on One spot.** *HMS Illustrious*

Above:

Illustrious **anchored in Falklands waters after the ship's Sea Harriers had been relieved on air defence duties by**

RAF Phantoms which were to be based permanently at RAF Stanley ashore. In the postwar navy, it is very unusual, if not unique, for a major warship to have a shake-down while on passage to and actually at a very 'hot' spot — Argentine forces had not confirmed the cessation of hostilities. *HMS Heron*

Above right:

On its way home from the Falklands *Illustrious* **paid an extremely successful visit to Fort Lauderdale. The carrier is pictured entering port with an extremely impressive deck park of two Sea King AEWs, nine Sea Harriers, six Sea King HAS5s and a Wessex HU5.** *HMS Illustrious*

the problem area. By the time the helicopters returned to Culdrose, their base, in December they had accumulated over 350 hours of flying time in test. It was on 21 October that the air defence responsibility was passed to the Royal Air Force whose Phantom fighters had been given a special runway extension at Port Stanley airfield by the Royal Engineers.

Accompanied by *Amazon* — a Type 21 frigate — and the tanker RFA *Brambleleaf*, *Illustrious* departed the FIPZ in 'style' with a fly past of 21 aircraft. The new carrier's cruising was not completed however, because it was tasked with courtesy visits to the East Coast of the United States, starting with Roosevelt Roads, Puerto Rico on 5 November. The crew had four days of well earned rest and recreation after 95 days at sea.

After three days at sea again, *Illustrious* entered Fort Lauderdale harbour to the delight of the American general public and with due ceremony. During the stay, 10,000 people visited the ship, but it was time to head home. After a call at Philadelphia Naval Yard, the ship departed for Portsmouth on 29 November. The aircraft departed on 6 December — with the Sea Harriers of 809 Squadron going back to RNAS Yeovilton to disband and the Sea Kings to RNAS Culdrose. After berthing at Portsmouth on

8 December, the ship spent some time in dockyard hands being inspected and repaired.

On 30 March 1983 *Illustrious* was re-dedicated by HRH Princess Margaret in lieu of the fully fledged commissioning ceremony which the ship had seemed to have missed in the rush to make it ready for operational duties. Other visits in 1983 included a visit to Tyneside to thank the people there for their hospitality, and Lisbon in May and June, before the major NATO exercise, 'Ocean Safari'.

The ship was again in the news during 7 to 17 June, because one Sea Harrier, from 801 Squadron, then embarked in *Illustrious*, became lost on exercise and made an emergency landing on a Spanish coaster, *Alraigo*. Although the aircraft was saved from destruction by very cool-headed flying the pilot, Sub Lt Ian Watson, was perhaps not the most popular pilot in the FAA for a few days! Eventually, the Sea Harrier was handed over by the Spanish crew at Tenerife and returned to the UK in *British Tay*. Watson was however flying again from *Illustrious* before his aircraft had actually been returned! After a visit to France and a change of Captain, *Illustrious* returned for summer leave.

In the autumn, the ship sailed for the Mediterranean where the Near East was suffering rather from the civil war situation in Lebanon, so

it was perhaps not surprising that at one time the RN had *Hermes*, *Illustrious* and *Invincible* operating in the same sea. The former two ships were taking part in the NATO exercise 'Display Determination' and the latter was en route for the Far Eastern deployment.

Above:
The return to Portsmouth for *Illustrious* in December 1982 — the air group had flown off to its bases at Culdrose and Yeovilton on passage up the Channel.
Mike Lennon

HMS *Ark Royal*

The third member of the 'Invincible' class has probably the most famous name in Royal Naval history. *Ark Royal* has been a name associated with aircraft carriers since December 1914 when an old steam collier was taken over by the RN for war service as a seaplane carrier — in fact, this was the first 'carrier of aircraft'. It served with distinction at the Dardanelles (1915) and later continued service during World War 1, mainly in the Mediterranean. When the Admiralty confirmed the requirement for a flat-top aircraft carrier in 1934, the seaplane carrier was rechristened *Pegasus*. It is sometimes considered unlucky to change the name of a warship, but *Pegasus* continued to serve until broken up in 1950.

The true aircraft carrier was launched in 1937 and was immediately hailed as a warship ahead of its time. *Ark Royal* became famous during the opening gambits of World War 2, not least because the Germans continually claimed to have sunk it. The ship gained Battle Honours for Norway (1940), Spartivento (1940), the Mediterranean (1940-41), the operations against the German pocket battleship *Bismarck* (1941) and finally for the Malta Convoys (1941). It was during one of these convoys that the third *Ark Royal* was torpedoed by a German submarine; the great carrier later sank off Gibraltar despite gallant attempts to save it.

In 1945 the names of two aircraft carriers then building for the Royal Navy were changed to names of carriers lost in combat. One was *Ark Royal* (ex-*Irresistible*). As much as anything, this name change was due to the efforts of the city of Leeds which collected funds for another carrier when their adopted *Ark* sank in November 1941. It was not until 1950 that the next *Ark Royal* was launched, by HM Queen Elizabeth (now the Queen Mother), who became patron of the ship and is reported to have dearly loved 'her' ship. The new *Ark* was commissioned in the Royal Navy in 1955 and served until 1978, when it decommissioned as the last British conventional fixed-wing aircraft carrier. During the intervening years, *Ark* became famous, not only because of the television series *Sailor* but also because of the several faults it developed. Nevertheless, the ship was the only carrier chosen for conversion to operate the McDonnell-Douglas F-4K Phantom FG1. Although there were many attempts to save it, the breaker's yard was the final destination in 1980.

8 'Orient Express'

In the four years from the decommissioning of the last conventional fixed-wing aircraft carrier, *Ark Royal*, to the beginning of the Falklands conflict, the light aircraft carrier *Invincible* has won the heart and affection of the British public. It was not until the Far Eastern group deployment, known as 'Orient Express', that the Royal Navy realised in how much esteem the ship is held by other nations, particularly Australia and New Zealand.

The Australian reaction is not surprising because in 1981 the British Government offered to sell *Invincible* to the Royal Australian Navy for the 'rock-bottom' price of £150million; in addition, Australia would purchase additional Sea King helicopters, which it already had flying from its ageing existing aircraft carrier, *Melbourne*. A new order for British Aerospace Sea Harrier fighters was also expected. The scheme was

Below:
***Invincible* had been selected to lead a group deployment to Australasia in 1982; it was to be the ship's last deployment before being handed over to the Royal Australian Navy. The Argentine invasion of the Falkland Islands intervened, delaying the deployment to the southern hemisphere. For 'Orient Express' *Invincible* was the leader and would remain very much a warship of the Royal Navy. With its ship's company in whites, *Invincible* wears the flag of Rear Adm Black, Flag Officer First Flotilla.** *HMS Invincible*

Above:
The 'Orient Express' Far East group deployment — *Invincible* **in the centre, with the three Royal Fleet Auxiliaries** *Appleleaf, Olmeda* **and** *Regent* **to the starboard side. To the carrier's port are the escort frigates** *Rothesay, Aurora* **and** *Achilles*.
Fleet Photographic Unit

greeted with mixed feelings in the UK, mainly because of the close relationship perceived by the general public, between the ship, the Sea King squadron (820) and HRH Prince Andrew, who was a Sub Lieutenant pilot on Sea Kings from 1981 until 1983; his service was to include operations with the carrier and squadron in the South Atlantic.

Others felt that the taxpayer would not be getting a good deal from the Australian sale, because the carrier had cost nearer £200million to build and was now fitted with several new systems. However, there was a school of thought that the British Government was right to sell the ship to Australia as long as a replacement was ordered from British Shipbuilders. The advantage of a fourth ship, even if the cost was more than £150million, which it undoubtedly would have been, was related to the advances and improvements which could be made in the ship, following early experience with *Invincible*. This was particulary true of the after end accommodation which is thought to have been added to the original design when it was at the CCH stage.

In the event, the RAN was not to receive a replacement aircraft carrier, even though the name 'Australia' had already been allocated for the ship. The Canberra government of Prime Minister Frazer was looking for a way out of the deal because so many defence dollars were tied up in the scheme. Australia is a large country with a small population (about 14.7million in 1983), and although the coastline is long there is no perceived threat to the country's sovereignty. A replacement carrier could only be afforded if it came second-hand and then it would still drastically effect the RAN's budget for several years after the purchase date. In addition, as during the 'Carrier Crisis' in the UK during the 1960s, the air force in Australia was lobbying for more aircraft and was claiming that they could carry out the roles of the air group planned for 'HMAS *Australia*'.

The result has been the winding down of the Royal Australian Navy's Fleet Air Arm, which by July 1984 was down to 30 helicopters and no fixed-wing aircraft. This has been a blessing in disguise for the Royal Navy however because a number of well-trained pilots have transferred to the RN's Fleet Air Arm and serve in carrier and ship-related flying posts. In addition, a certain shortfall in the numbers of RN Observers is being filled from the Australian source.

The 'Orient Express' deployment followed closely behind a smaller and short exercise to the Caribbean in early 1983 — 'Caribtrain '83'. The purpose of this deployment was to test the weapons systems in the ship, including the air group on the American range facilities. Flag showing was also carried out in several locations including mainland USA and several Caribbean islands.

The Far Eastern deployment was a far longer and more important deployment, being the most important 'out of area' operation carried out by the RN since the withdrawal from the Singapore base in October 1971. Although since 1971 there have been a number of deployments, including the very important visit to the People's Republic of China in 1980, it is only recently that the UK has been seriously considering its role outside the NATO area of the North Atlantic (as far south as the Tropic of Cancer). The South Atlantic conflict and the permanent squadron of British warships in the Gulf of Oman/Arabian Sea area has shown the world that the RN is still a global force, but 'Orient Express' was an effective way of furthering British Government policy objectives by demonstrating an effective presence in the Indian Ocean, South China Sea and surrounding areas.

The 'Orient Express' group, under the overall command of Rear Adm Jeremy Black DSO, MBE — known as 'JJ' to the men he took to the Falklands and back when Captain of *Invincible* — was led by *Invincible*. In addition, the frigates *Aurora*, *Achilles*, *Rothesay* and *Andromeda* accompanied the group at various times during the first half of the deployment, from the departure from Portsmouth in September 1983 until Christmas in Australia. The group was supported by the Royal Fleet Auxiliaries *Regent* (a fleet replenishment ship), *Olmeda* (fleet tanker) and *Appleleaf* (supply tanker). After Christmas *Invincible* was supposed to call at Sydney, and join the guided missile destroyer *Glamorgan* together with the Type 22 frigate *Brazen* in the Indian Ocean.

Invincible carried the five Sea Harrier FRS1 fighters of 801 Squadron, under Lt Cdr Tony Ogilvy, and the eight Sea King HAS5 helicopters of 820 Squadron, under the command of Lt Cdr Gerry Hunt. The Captain of *Invincible*, Capt The Honourable Nicholas Hill-Norton, was disappointed that a flight of two Sea King AEW helicopters could not be made available to the ship, nor could an additional three Sea Harriers and two Sea King anti-submarine helicopters, making the total air group 20 aircraft strong. This

Above:

In the Mediterranean, off the North African coast, is a naval anchorage for the Soviet Mediterranean Squadron and elements of the Black Sea Fleet. Here, *Invincible's* Sea Kings and the frigate *Achilles* investigated the new 'Krasina' class cruiser *Slava* (126), seen for the first time in Western waters. *Fleet Photographic Unit*

Left:

Sailing in the Indian Ocean after a joint exercises with French naval forces in the Red Sea area, *Invincible* enjoys a 'Sunday' routine, with its aircraft ranged on the flight deck for the next day's ceremonial port entry at Bombay. *RN*

Above right:

The French Admiral Le Febvre, commanding the French naval forces in the Indian Ocean and going by the title ALINDIEN, was welcomed aboard *Invincible* by Rear Adm Black and Capt Hill-Norton (right). *HMS Invincible*

would have given more scope for the joint exercises carried out as the 'Orient Express' group moved around the Far East. Initially there was a two week work-up period in the English Channel and Bay of Biscay, when the opportunity was taken to exercise with Royal Netherlands, Portuguese and Spanish Navy submarines.

After a short stop at Gibraltar, a brief encounter with some Soviet warships was made in one of their anchorages, some 15 miles off the Libyan coast. This was an opportunity to examine at close hand one of the latest Soviet warships, *Slava* of the 'Krasina' class. This vessel was built in the Black Sea and rates as a 12,000-ton cruiser. The 'Orient Express' group caught *Slava*, plus auxiliaries, on its first training cruise into the Mediterranean. Later in the voyage several other Soviet warships were seen and identified.

Although it was not possible for the group to exercise with the Egyptian Navy during the outward passage, the sights of ancient Egypt were seen by many from the ships involved.

Such a deployment as 'Orient Express' not only has to show the flag — or indicate to the nations visited that the UK has a continuing interest in promoting good relations with them, and has the capability and the will to do so, as the official statement said — but also to engage in mutual beneficial exercises. Certainly the large scale operations in which *Invincible*, together with *Aurora* and *Rothesay* plus the RFAs, engaged during passage in the Indian Ocean — and after Fremantle, the Southern Ocean — were designed

to test the RN/RAN co-operation and inter-operability. The RAN is perhaps the closest aligned navy to that of the UK, outside of NATO, and the opportunity of working with up to 12 Australian escort and fast patrol ships was judged to have been a great success, even if opinions differed as to who won the 'Tasmanian Mini-War'.

When *Invincible* passed through the Bass Straits, from the Great Australian Bight into the Tasman Sea, it hosted two Sea King HAS50 helicopters from 817 Squadron Royal Australian Navy, which had been, until 1981, embarked in the carrier *Melbourne*. The Sea Harriers also took the opportunity of flying air combat manoeuvring (ACM) sorties against RAN A-4G Skyhawks and RAAF Mirage IIIOE fighters.

The joint RAN/RN Sea King operations were aimed at identifying, classifying and 'attacking' the Australian 'Oberon' class conventional submarine *Ovens*. Both Sea King types were equipped with passive and active sonar locating systems, although of different manufacture, as were the helicopters' radars. Nevertheless, and although 820 Squadron does not operate MAD (magnetic anomaly detection)-equipped Sea Kings, the submarine was identified and attacked on numerous occasions. The British frigates were also able to work up their surface-launched ASW systems with Australian frigates and destroyers.

Smaller scale exercises had been conducted with the Singaporeans earlier in the deployment, when the RN warships and the patrol vessels of the South-East Asian nation were able to work together in anti-piracy, shipping identification and similar roles. Singapore's neighbour, Malaysia, was interested in buying both the Sea King helicopter (as carried by *Invincible*) and the Westland Lynx (embarked in *Andromeda*). The demonstration of British equipment was also one of the aims of the deployment — India being another potential user of certain British systems.

Although it seems unlikely that the Indian Navy will acquire an 'Invincible' class carrier, Rear Adm Black and Capt Hill-Norton were extremely

Below:

On passage, the 'Orient Express' group worked with the American forces, including the aircraft carrier *Ranger*, whose air group and escorts exercised with the British warships. *HMS Invincible*

Bottom:

Demonstrating Anglo-American co-operation in the Indian Ocean, a Sea Harrier FRS1 flown by Lt Cdr Dave Braithwaite launches from the *Ranger's* large flight deck. During mock air battles the Sea Harriers again showed that being subsonic does not means that they come off badly in air combat manoeuvring — usually it is the opposite. *RN*

pleased with the interest shown in the carrier's equipment by the Indian Navy delegations which visited the ship at Bombay, Goa and Cochin. India has ordered two batches of Sea Harrier from British Aerospace, including a second order of 13 as a result of the *Invincible* visit and a 'shop window' flying display.

Ships' companies are used to their programmes changing rapidly, but the changes to the second — 1984 — part of *Invincible's* deployment took even the most experienced by surprise. The planned visit to Hong Kong was cancelled with the carrier going instead to Singapore for a second visit and an opportunity to dry dock the carrier in order that engineers could examine the faulty shaft bearing which had been causing problems since the ship left Fremantle. The problem with the shaft was exaggerated in the British and Australian press which had almost written the ship off, but the fact is that it could still operate aircraft throughout the flying 'envelope'. In peacetime, however, especially when a ship is far from home, it is prudent to ensure that any faults are rectified as soon as possible. The matter was complicated by the rather ridiculous stance of the Australian Government which has a ban on nuclear weapons on Australian soil — the dry dock was considered 'soil'. Because Sydney has the most important dry-docking facilities in that part of the world, and because *Invincible* was in southern waters, it was natural for the British Government to request the use of the facilities to repair the bearing fault. It is official policy for a ship's commanding officer never to admit or deny that nuclear weapons are being carried in his ship. A problem therefore arose because the Australian Government was rather simplistic in its approach to the problem and because Capt Hill-Norton would not deny the presence of nuclear devices for his aircraft: the Australians said that such devices must be present, as part of the ship's armament, but that there was nothing to stop

Above left:
Groundcrew from 801 Squadron wait to clean the engine intakes and other parts of the Sea Harrier after a sortie during the 'Mini-War' fought by the 'Orient Express' group with Australian units. The aircraft, OOO, is the 'mount' of Lt Cdr Tony Ogilvy, the CO of 801 Squadron. *Author*

Left:
Waiting on deck for *Regent's* Wessex 5 to clear, one of two Royal Australian Navy Sea King helicopters which took up temporary residence in *Invincible* in November 1983. *Author*

Invincible tying up alongside at the port — that was not apparently Australian soil.

For one of Britain's oldest allies to take such a confusing stance is really quite unbelievable, and not surprisingly the British Government asked the Singaporeans if their dry docking facilities could be used. Whilst these negotiations were in progress, the Australians came back to the Ministry of Defence in London and said, in so many words, that their position had been misunderstood — they did not allow a ship to go into dry dock with 'significant explosives' aboard. The British feeling was, apparently, that they were already engaged in the final stages of the agreement to put Invincible into Sembawang, the former British naval base complex in Singapore, and therefore could not take advantage of the Canberra Government's change of heart.

The crisis in the Lebanon and in the Arabian Gulf further complicated the programme of the 'Orient Express' deployment, with the scaling down of joint exercises with RN frigates and destroyers coming out from the UK to join Invincible. Glamorgan and Brazen were diverted to the Lebanese coast to join the amphibious

warfare ship Intrepid which was standing by to take off British troops and civilians in the event of civil war. Nevertheless, the whole 'Orient Express' operation was a tremendous success — proving that prolonged 'out of area' operations were possible and that there is considerable merit in deploying naval forces around the world.

After returning to UK waters, Invincible underwent a docking and essential defects (DED) programme at Plymouth, before embarking an enlarged helicopter force for anti-submarine warfare exercises and trials in the Atlantic. The ship's half-life refit is scheduled for 1985-6.

Below:
En route to Australia from Singapore, *Invincible* completes a RAS — replenishment at sea — from *Regent*, the Group's solid stores carrier and replenishment ship. *HMS Invincible*

Bottom:
Said by all to be one of the 'best runs ashore', Fremantle in Western Australia was very welcoming to *Invincible*, *Aurora*, *Rothesay* and *Regent*. On deck are ranged three of 801 Squadron's Sea Harriers and seven of 820 Squadron's Sea Kings; the other two Sea Harriers and the remaining Sea King were displayed in the hangar for visitors. *HMS Invincible*

Appendices

1   'Invincible' Class Specifications

	Invincible	Illustrious	Ark Royal
Pennant number	R05	R06	R09
Flight deck code	N	L	R
Ordered	17/4/73	14/5/76	1/12/78
Laid down	20/7/73	7/10/76	14/12/78
Launched	3/5/77	1/12/78	2/6/81
Completed	19/3/80	18/6/82	1984
Commissioned	11/7/80	30/3/83*	1985
Standard displacement (tons)	16,000	16,000	16,000
Full displacement, as built (tons)	19,500	19,810	20,000
Full displacement, 1984 (tons)	19,810	19,810	20,000
Length (overall)	677ft	677ft	677ft
Length (waterline)	632ft	632ft	632ft
Beam	90ft	90ft	90ft
Draught	24ft	24ft	25ft
Range	5,000nm at 18kt		
Speed (max)	28kt		
Speed (cruising)	18kt		
Complement (ship's company)	131 officers and 869 ratings		
Complement (air group)	65 officers and 135 ratings		
Builders	Vickers	Swan Hunter	Swan Hunter

Below:
Illustrious in January 1982, with its engine 'alight', is manoeuvred passed the shape of Ark Royal fitting out on the Tyne. Although the ship's shell would seem complete, much of the internals would have normally taken 12 months or more to complete. The yard completed the work in less than half the time.
Swan Hunters

*Because of the Falklands conflict, *Illustrious* was accepted into service by the Royal Navy on 21 June 1982 in a unique service at sea.

2 Weaponry Specifications

British Aerospace Sea Harrier FRS1

Wing span: 25.25ft (7.70m)
Overall length: 47.58ft (14.5m)
Length (nose folded): 42.25ft (12.88m)
Height: 12.14ft (3.7m)
Underwing load: 5,000lb (2,268kg)
Fuel capacity: 5,020lb (2,277kg)
All-up weight: 23,000lb (10,433kg)
Engine: 1 Rolls-Royce Pegasus 104 turbofan

Maximum speed: 625kt (1,158km/hr)
Cruising speed: 485kt (898km/hr)

Below:
801 Squadron took its Sea Harriers aboard *Invincible* in 1981 and the Squadron was still embarked when the ship deployed to the South Atlantic a year later. Lt Cdr Nigel 'Sharkey' Ward is the pilot of this aircraft; at the time, he was the Squadron CO. *801 Squadron/N. Thomas*

Westland Sea King

This data applies equally to the Sea King HAS5 (anti-submarine version) and the Sea King AEW (airborne early warning version), except where otherwise noted:

Overall length, rotor turning: 72.67ft (22.15m)
Overall length, fuselage: 55.81ft (17.01m)
Overall height, rotors turning: 16.83ft (5.13m)
Overall width, rotor/tail folded: 16.40ft (5.00m)
Main rotor diameter: 62.00ft (18.90m)
Main cargo door size: 5.67ft×5.00ft (1.72m×1.52m)
Max all-up weight (AEW): 21,400lb (9,714kg)
Max all-up weight (Mk 5): 21,000lb (9,526kg)
Max all-up weight (Mk 2): 20,500lb (9,298kg)
Max cruising speed: 112kt (207km/hr)
Radius of action (AEW, 3.75-hour endurance): 20nm (37km)
Range (ASW): 664nm (1,230km)
Endurance (ASW sortie): 5 hours
Endurance (SAR sortie): 3.75 hours (including rescue hover)
Engines: 2 Rolls-Royce Gnome H1400-1 (1,660shp)
Weapons fit (ASW): 4 DC11 depth bombs; 4 Mk 44, Mk 46 or Stingray torpedoes; nuclear depth
   bombs; 1 general purpose machine gun (door)

NB: It is considered likely that the Sea King HAS5 will be fitted with the British Aerospace Dynamics Group P3T Sea Eagle anti-ship missile system in the near future.

Sonar equipment (ASW): 1 Type 195 dunking active/passive set (to be replaced by HISOS in 1988)
Radar equipment (ASW): MEL Sea Searcher (50nm/93km)
Processors (ASW): Marconi LAPADS AQS-901
Other ASW equipment: Mini-Jezebel Sonobuoys; ASQ-501 Magnetic Anomaly Detector (MAD)
Radar (AEW): ARI 5955 retained but primarily Thorn-EMI Searchwater set

EH101 Future Helicopter Design

Overall length, rotor turning: Data not available
Overall height, rotor turning: 21.30ft (6.50m)
Main rotor diameter: 61.00ft (18.90m)*
Maximum cruising speed: 180kt (333km/hr)
Range: 1,100nm (2,035km)
Endurance: 5 hours
Engines: 3 General Electric GE-T700-401
Weapons fit: 1 Exocet AM39 or Sea Eagle missile; 2 Oto-Melara Marte missiles (anti-ship); 2 Mk 46, Stingray or similar torpedoes; nuclear depth bombs
Sensor equipment: 1 Plessey/Marconi HISOS-2 Guillemot
Processors: 1 Marconi LAPADS AQS-903
Radar: 1 Ferranti Blue Kestrel, or similar
Other ASW equipment: 1 ASQ-501 Magnetic Anomaly Detector; Mini-Jezebel sonobuoys
Maximum all-up weight: 28,665lb (13,000kg)
*Same as present Sea King

Sea Dart GWS30 Missile System

Length: 14.3ft (4.36m)
Body diameter: 1.38ft (0.42m)
Rate of fire: By salvo (classified)
Velocity: Mach 2.5

Missile weight: 1,213lb (550kg)
Warhead charge: HE (weight classified)
Range: 16nm (30km) in anti-ship role; 20nm (37km) in anti-air role

Type GAM-BO1 20mm Oerlikon Cannon

Rate of fire: 1,000 rounds per minute
Muzzle velocity: 1,100-1,200m/sec (3,610-3,937ft/sec)
Mass of gun (without ammunition): 926lb (420kg)

Mass of gun (with 200 rounds): 1,102lb (500kg)
Types of ammunition: HEI (high explosive incendiary); SAPHEI (semi-armour piercing high explosive incendiary); AP (armour piercing)

HEI ammunition is relatively thick-walled giving high fragmentation with both blast and incendiary effects which are lethal against small craft and most aircraft.

SAPHEI ammunition is made of high-grade steel for penetration into the target where the delayed action base fuze gives blast, fragmentation and incendiary effects. It is particularly good against aircraft.

CIWS Mk 15 20mm Vulcan Phalanx

Rate of fire: 3,000 rounds per minute
Muzzle velocity: Classified data
Mass of gun (without ammunition): 2,205lb (1,000kg)

Traverse arc: 360deg
Elevation arc: −35deg to +90deg
Range: Approx 1nm (1.9km)
Types of ammunition: HEI, SAPHEI, AP